Hydrometeorological Extreme Events and Public Health

Other Hydrometeorological Extreme Events Titles

Hydrometeorological Extreme Events and Public Health

Edited by

Franziska Matthies-Wiesler

Institute of Epidemiology
Helmholtz Zentrum München
German Research Center for Environmental Health
Ingolstädter Landstrasse 1
D-85764 Neuherberg
Germany

Philippe Quevauviller

Vrije Universiteit Brussel (VUB)
Dept. of Hydrology and Hydrological Engineering
c/o Drève de Nivelles, 143
B-1150 Brussels
Belgium

WILEY Blackwell

The right of Franziska Matthies-Wiesler and Philippe Quevauviller to be identified as the authors of the editorial material in this work has been asserted in accordance with law.

Registered Offices
John Wiley & Sons, Inc., 111 River Street, Hoboken, NJ 07030, USA
John Wiley & Sons Ltd, The Atrium, Southern Gate, Chichester, West Sussex, PO19 8SQ, UK

Editorial Office
9600 Garsington Road, Oxford, OX4 2DQ, UK

For details of our global editorial offices, customer services, and more information about Wiley products visit us at www.wiley.com.

Wiley also publishes its books in a variety of electronic formats and by print-on-demand. Some content that appears in standard print versions of this book may not be available in other formats.

Library of Congress Cataloging-in-Publication Data
Names: Matthies, Franziska, editor. | Quevauviller, Ph., editor.
Title: Hydrometeorological extreme events and public health / edited by Franziska Matthies-Wiesler, Philippe Quevauviller.
Description: Hoboken, NJ : John Wiley & Sons, Ltd, 2022. | Includes bibliographical references and index.
Identifiers: LCCN 2021045162 (print) | LCCN 2021045163 (ebook) | ISBN 9781119259305 (hardback) | ISBN 9781119259343 (pdf) | ISBN 9781119259251 (epub) | ISBN 9781119259350 (ebook)
Subjects: LCSH: Medical climatology. | Human beings--Effect of climate on. | Public health.
Classification: LCC RA793 .H93 2022 (print) | LCC RA793 (ebook) | DDC 616.9/88--dc23/eng/20211220
LC record available at https://lccn.loc.gov/2021045162
LC ebook record available at https://lccn.loc.gov/2021045163

Cover Image: © Marc Bruxelle/Shutterstock
Cover Design by Wiley

Set in 10/12pt Times by Integra Software Services Pvt. Ltd, Pondicherry, India
Printed and bound by CPI Group (UK) Ltd, Croydon, CR0 4YY

C9781119259305_110322

Contents

List of Contributors

Ian Clark
Senior Adviser to the European Commission

Máire Connolly
Professor of Global Health and Development, National University of Ireland Galway (NUIG) Galway, Ireland

Paul Hudson
Department for Environment and Geography, University of York, United Kingdom

Demetrio Innocenti
University of Antwerp, Belgium

Owen Landeg
National Institute for Health Protection, London School of Hygiene and Tropical Medicine, United Kingdom

Franziska Matthies-Wiesler
Institute of Epidemiology, Helmholtz Zentrum München, German Research Center for Environmental Health, Germany

Lidia Mayner
Torrens Resilience Institute, College of Nursing and Health Sciences, Flinders University. Adelaide, South Australia

Virginia Murray
Head of Global Disaster Risk Reduction and a COVID-19 Senior Public Health Advisor, Public Health, England

Antonio Navarra
Universita' di Bologna, Bologna, Italy

Philippe Quevauviller
Vrije Universiteit, Brussels, Belgium

Gerardo Sanchez Martinez
The UNEP DTU Partnership, Copenhagen, Denmark

Enrico Scoccimarro
Fondazione Centro Euro-Mediterraneo sui Cambiamento Climatici, Bologna, Italy

Alistair Woodward
University of Auckland, New Zealand

Editors

Franziska Matthies-Wiesler is a Senior Researcher at the Institute of Epidemiology at the German Research Centre for Environmental Health, Helmholtz Zentrum München, since 2019. She also works as a free-lance consultant in the area of environment and health following on from her position at the European Centre for Environment and Health of the WHO Regional Office for Europe.

Since 2001, her major field of expertise is global climate change and human health, in research as well as in policy analysis, advice and advocacy, e.g. for the German Advisory Council on Global Change (WBGU), the Tyndall Centre for Climate Change Research, the WHO Regional Office for Europe and the European Commission. Her main focus includes public health preparedness and response to extreme weather events, health co-benefits of climate change adaptation and mitigation. More recently she has broadened her field to include holistic concepts such as health in the context of the sustainable development goals (SDGs), the planetary health concept and transformational change. She has contributed to research projects co-funded by the European Commission and within the Horizon Europe programme. Since March 2021 she is board member of the German Alliance Climate Change and Health (KLUG) and since 2016 of IEH Consulting, UK.

Philippe Quevauviller has been researcher in chemical oceanography in the years 1983-1989 firstly at the University of Bordeaux, then at the Portuguese Environment State Secretary, and as a post-doc researcher at the Dutch Ministry for Public Works (Rijskwaterstaat). From this research work, he gained two PhD in oceanography (1987) and chemistry (1990) at the University of Bordeaux, and later an HDR (Diplôme d'Habilitation à Diriger des Recherches) in chemistry at the University of Pau, France (1999). He joined the European Commission in 1989 in the chemical metrology sector (quality assurance of environmental analyses) and moved in 2002 to the policy sector where he has been in charge of a new EU Directive on groundwater protection against pollution. He moved back to the EC DG Research in 2008, as a research Programme Officer in the field of climate change impacts on the aquatic environment and on hydrometeorological hazards. Since 2013, he moved to another area of work, namely Security Research, at the EC DG Home Affairs where he is Research Programming and Policy Officer in the fields of Disaster Risk Management and in charge of the development of a Community of Users on Secure, Safe and Resilient Societies.

Besides his EC career, Philippe Quevauviller has been active in academic and scientific developments as Associate Professor at the Free University of Brussels(VUB) and promoter of Master theses in an international Master on water Engineering (IUPWARE programme). It is under this function that he is acting as Series Editor of the *Hydrometeorological Extreme Events* Series for Wiley. He has published (as author and co-author) more than 250 scientific and policy publications in the international literature and 20 books; he has also coordinated a book Series for Wiley on *"Water Quality Measurements"* which resulted in 10 books published between 2000 and 2011.

Series Preface

The increasing frequency and severity of hydrometeorological extreme events are reported in many studies and surveys, including the 6th IPCC Assessment Report. This report and other sources highlight the increasing probability that these events are partly driven by climate change, while other causes are linked to the increased exposure and vulnerability of societies in exposed areas (which are not only due to climate change but also to mismanagement of risks and 'lost memories' about them). Efforts are on-going to enhance today's forecasting, prediction and early warning capabilities in order to improve the assessment of vulnerability and risks and develop adequate prevention, mitigation and preparedness measures.

The Book Series on 'Hydrometeorological Extreme Events' has the ambition to gather available knowledge in this area, taking stock of research and policy developments at international level. While individual publications exist on specific hazards, the proposed series is the first of its kind to propose an enlarged coverage of various extreme events that are generally studied by different (not necessarily interconnected) research teams.

The Series encompasses several volumes dealing with various aspects of hydrometeorological extreme events, primarily discussing science-policy interfacing issues, and developing specific discussions about floods, coastal storms (including storm surges), droughts, resilience and adaptation, governance, health impacts, etc. While the books are looking at the crisis management cycle as a whole, the focus of the discussions is generally oriented towards the knowledge base of the different events, prevention and preparedness, early warning and improved prediction systems.

The involvement of internationally renowned scientists (from different horizons and disciplines) behind the knowledge base of hydrometeorological events makes this series unique in this respect. The overall series will provide a multidisciplinary description of various scientific and policy features concerning hydrometeorological extreme events, as written by authors from different countries, making it a truly international book series.

The Series so far is made of four volumes, an introductory one, a second volume dealing with coastal storms, a third volume on droughts and a fourth volume about governance aspects. The 'Health' volume is the fifth book of this Series; it has been written by experts in the field, covering various horizons and (policy and scientific) views. It offers the reader an overview of scientific knowledge about climate change impacts on health, with discussions on specific climate/weather extreme events and their effects, disaster risk reduction actions (in particular preparedness and response) related to health, and economic aspects of health-related costs of hydrometeorogical extreme events.

Ph. Quevauviller
Series Editor

Foreword

Extreme weather events have impacted human society over millennia and have changed the course of history. Climate change is increasing the risk of extreme events, which can cause loss of life, injury or other health impacts, property damage, loss of livelihoods and services, social and economic disruption, or environmental damage. *Hydrometeorological Extreme Events and Health* addresses the key challenge of managing the health impacts of these weather events.

The hydrometeorological hazards with the highest impacts include flooding, hurricanes, cyclones, typhoons and droughts. The health effects of these events include deaths due to direct trauma, injuries, increased incidence of infectious diseases, and mental health impacts due to loss of relatives, housing and livelihoods.

Floods affect countries worldwide and cause greater mortality and infrastructural damage than any other type of hazard. Flooding can disrupt water supply and sewage systems, and can cause water waste to overflow leading to water-borne disease outbreaks. Severe flooding in Germany and Belgium in 2021 led to significant loss of life and widespread devastation. Large numbers of healthcare facilities shut down due to infrastructural damage, leaving affected communities with poor healthcare coverage. Flash flooding events occur in the Mediterranean region including the southeast of France, northern Italy, Catalonia and Corsica with high levels of devastation but preparedness is better with greater awareness of risk amongst populations and higher levels of infrastructural resilience. Flooding in the southern states of the USA in 2021 during the COVID-19 pandemic led to a major surge in mortality and morbidity, which required deployment of military teams to provide medical care.

Hurricanes, cyclones and typhoons are severe storms that form over tropical water, which can cause widespread infrastructural damage and population displacement with health consequences. Droughts are slow onset events that can cause large agro-ecological damage, socio-economic effects, and food insecurity with negative health impacts such as malnutrition.

Over the course of 10 years as Coordinator of the WHO Disease Control in Emergencies Programme at WHO/HQ working on risk assessments and disaster response, there have been major advances in the forecasting, prediction and early warning of extreme weather events with assessments of vulnerability and risks leading to prevention, mitigation and preparedness measures. However, the public health measures required to improve healthcare resilience for extreme weather events has not received the same level of investment as other sectors. Disaster preparedness scenarios of high impact events to assess the capacity and capability of countries to respond to crises such as epidemics and major weather events to include disaster risk reduction are a core part of preparedness measures. This needs to be informed by up-to-date research findings and policy developments.

Hydrometeorological Extreme Events and Health is the fifth volume of a series which gathers scientific and policy related knowledge on climate-related extreme events. Climate change has led to greater investment in research on the transfer of water and energy between land and the lower atmosphere in the combined scientific domains of meteorology and hydrology. The editors of the book have extensive experience in the areas of environment and health, public health preparedness and response, climate change, hydrology and disaster risk management. The contributors are internationally recognized experts in their respective fields who have built up global networks collaborating on state-of-the-art research. The chapter authors bring an invaluable knowledge base in the areas of environmental sciences, epidemiology, public health and disaster risk reduction from their work in academia, government authorities, EU institutions and UN agencies. Their expertise is brilliantly captured in the chapters outlining a broad range of topics, including precipitation and temperature extremes in a changing climate, climate change and health, health effects of flooding, disaster risk reduction and health, preparedness and response in the context of climate change and the health costs of extreme events. This broad multi-disciplinary approach written by authors from different countries brings valuable perspectives on the urgent action needed to reduce the risk of extreme events and ensure better preparedness at local, national, EU and global levels.

As we move into the recovery phase of a global pandemic and assess our preparedness for future hazards including extreme weather events, it is important to ensure a broader focus incorporating the implementation of the UN Sustainable Development Goals (SDGs) by all countries and embracing the concept of planetary health. This book is a valuable compilation of the most recent scientific, research and policy developments in the field.

Máire Connolly
Professor of Global Health and Development
National University of Ireland Galway (NUIG)
Galway, Ireland

1

Introduction

Ian Clark
Former Head of unit of the European Commission

Since early 2020, due to COVID-19, the world has faced its most serious health crisis for a century with an enormous toll on human life together with severe effects on the health of the population and strains on health systems as well as on socio-economic well-being. More than 2 years since the first cases, the risks of the pandemic are still high. This health crisis has been a stark reminder of the dangers of insufficient preparation as most countries throughout the world were clearly not ready to face such a crisis and much of the response, certainly at the early stage, was ad hoc. At the same time, COVID-19 is a 'simple' crisis compared to the climate crisis, which is the most complex and intractable problem facing humankind. While recovery from Covid is a major challenge for the whole world and will take years, there are clear solutions, in particular through vaccination of a very large share of the population. These solutions are backed by a strong consensus at both policy and scientific levels which, however, is not the case for climate change. Meanwhile, the health threats due to climate change are increasing. The last decade was the hottest on record, with eight of the hottest years ever recorded. The health consequences of these increased temperatures include death and injury from extreme precipitation, tropical cyclones, heatwaves, floods, forest fires, as well as storm surges. The world can also expect the emergence and spread of infectious diseases and allergens linked to geographical shifts in vectors and pathogens. There will be additional challenges for the capacities of health systems due, for example, to the spread of previously unknown diseases from the Southern Hemisphere to the Northern Hemisphere.

This book is a timely publication to contribute to science and policy debates as we emerge from the Covid crisis and also as scientific input to underpin the upcoming COP26 deliberations and decisions.

2021 saw a new worldwide political impetus to reinforce climate action in support of health post-Covid building on the strong health provisions of the 2015 Paris agreement on climate change, as well as the 2015 Sendai Framework for Disaster Risk Reduction. The COP26 in Glasgow November 2021 was a major milestone for agreement on increased action. Already in April 2021 at the high-level summit to prepare COP26, a number of countries made renewed and updated pledges. Health and security also

Hydrometeorological Extreme Events and Public Health, First Edition. Edited by
Franziska Matthies-Wiesler and Philippe Quevauviller.
© 2022 John Wiley & Sons Ltd. Published 2022 by John Wiley & Sons Ltd.

featured at the summit – countries aim to scale up locally led solutions to climate vulnerability. Furthermore, in Europe, new policy developments in the EU aim to reinforce action to address the health challenge brought by climate change: a new Climate Adaptation Strategy launched in February 2021 proposes reinforced action to better understand climate-related risks to health and to increase capacity to reduce these risks. In addition, the revised Union Civil Protection Mechanism of May 2021 reinforces civil protection actions to help address climate change as well as health emergencies.

Over the past decade, a considerable body of scientific and policy literature has been developed analysing the impact of weather events on health but much less on long-term weather–climatic impacts. This publication aims to also contribute to the latter growing debate.

The book analyses and summarises the state of knowledge of climate change and certain specific extreme weather events for health, including the question of the health costs of extreme events and secondly addresses the international efforts since the early years of the century and particularly the agreements of 2015 to integrate health and climate action together.

Most of the extreme events mentioned above that cause health impacts are linked to precipitation and temperature, which are the subjects of Chapter 2. Based on assessments prepared using General Circulation Models, their performance in assessing potential health impact results for three periods is analysed: 1996–2014, 1996–2005 and forecasts for 2061–2100. It discusses the advances in modelling as well as the inaccuracies.

Chapter 3 focuses on the state of scientific knowledge about climate change and health, starting with the most recent IPCC assessment report from 2014 completed by summaries of more recent studies.

Flooding is the most frequent disaster worldwide and the health consequences are the subject of Chapter 4, looking at both the short- and long-term impact both on population well-being as well as the potential impact on health systems.

Chapter 5 includes a comparative analysis of the three 2015 international agreements – the Paris climate agreement, the Agenda 2030 agreement on Sustainable Development and the Sendai Framework for Disaster Risk Reduction – both their new and reinforced actions to address human health and well-being and the links between the three agreements. According to the World Health Organization (WHO), the Paris climate agreement is potentially the strongest international health agreement of this century. Building on this comparison, Chapter 6 analyses the international health agreements developed by WHO and how these have evolved in recent times to address more specifically climate and extreme weather impacts. The second part of Chapter 6 outlines specific diseases associated with the impacts of extreme hydrometeorological events. Finally, Chapter 7 presents the results of a wide body of academic work estimating health costs of extreme weather events, a subject neglected in most economic impact assessments of such events. Finally, the book identifies gaps and areas for additional study and calls for more attention in decision-making to account for the significant health costs.

2

Precipitation and Temperature Extremes in a Changing Climate

Enrico Scoccimarro[1] and Antonio Navarra[2]

[1] *Fondazione Centro Euro-Mediterraneo sui Cambiamento Climatici, Bologna, Italy*
[2] *Universita' di Bologna, Bologna, Italy*

2.1   Introduction

Episodes of large societal and economic relevance are caused by extreme weather events (Parry et al. 2007; Peterson et al. 2008). When estimating the impacts of climate change, the potential changes in climate variability and hence extreme events present a prime concern.

The distinction between *extreme weather events* and *extreme climate events* is not clear, but mainly relates to the timescale they refer to. An extreme weather event is associated with changes in weather conditions for a short period, from less than a day to a few weeks. An extreme climate event occurs over a longer period. It can be the accumulation of several weather events (not only extreme ones), such as a long period of days associated with low precipitation leading to a drought period (Seneviratne et al. 2012).

Different extreme events affect our society over different regions of the planet and people are subject to different kinds of weather extremes, with some places also subject to more than one kind of event. Since 1980 in the United States, there have been more than 10,000 deaths and in excess of 1.5 trillion USD in damage due to extreme weather events (NCEI, 2019; Smith and Matthews, 2015). The total reported economic losses caused by weather extremes for the same period over Europe amounted to approximately 453 billions of euros (EEA 2019). Many climate extremes are the result of natural climate variability (including phenomena such as El Niño), and natural decadal or multi-decadal variations in the climate provide the backdrop for anthropogenic

Hydrometeorological Extreme Events and Public Health, First Edition. Edited by
Franziska Matthies-Wiesler and Philippe Quevauviller.
© 2022 John Wiley & Sons Ltd. Published 2022 by John Wiley & Sons Ltd.

climate changes (Seneviratne et al. 2012). Even if there were no anthropogenic changes in climate, a wide variety of natural weather and climate extremes would still occur. Heatwaves, extreme precipitation, droughts, floods, storm surges and tropical cyclones are the most common extreme events providing risks and causing damage to our society (Bell et al. 2018). The evaluation of their impact is region dependent and many research gaps must still be addressed to improve the resilience of public health to such events. Most of these events are linked to two parameters – temperature and precipitation – that are considered reasonably well represented by state-of-the-art general circulation models (GCMs). A good characterisation of the precipitation event probability distribution helps not only to identify extreme precipitation conditions but also to assess drought and floods, together with precipitation patterns associated with tropical cyclones (Scoccimarro et al. 2014). On the other hand, a good representation of the temperature events distribution gives support not only to the identification of extreme daily temperature events but also tropical nights and heatwaves (Russo et al. 2014; Silliman et al. 2008; Zampieri et al. 2016).

Because of the temperature and precipitation roles in affecting human health and natural systems, it is important to assess potential changes in the tails of their distribution of events (the rarest conditions and/or events) under different climatic conditions. Changes in extremes can be linked to changes in the mean, variance or shape of probability distributions. However, all of these changes in extremes can also be directly related to changes in the mean climate, because mean future conditions in some variables are projected to lie within the tails of the distributions of present-day conditions. It is well known that in several cases the changes in extremes scale closely with changes in the mean (e.g., Griffiths et al. 2005), but there are also exceptions (such as short duration precipitation and temperature at high latitudes) highlighting changes in the shape of the probability distributions of weather and climate variables when focusing on future projections (Ballester et al. 2010; Brown et al. 2008; Della-Marta et al. 2007; Kharin et al. 2007; Orlowsky and Seneviratne 2011; Scoccimarro et al. 2013, 2016).

Modelling advances now provide the opportunity of utilising global GCMs for projections of extreme temperature and precipitation indicators. Most of the available scientific literature on climate extremes is based on the use of extreme indices, which can either be based on the probability of occurrence of given quantities or on threshold exceedance. Typical indices that are seen in the scientific literature are based on percentile analysis, where the 90th or 99th percentile, generally defined for given time frames (in the present chapter the time frequency considered is daily), are used to identify intense and extreme conditions. For example, the 99th percentile of a time series of daily precipitation values for a 30-year period, considering only the June to August (JJA) period (92 days × 30 years = 2760 values), correspond to the precipitation value – within the sample of 2760 numbers – exceeded only 27 times. Such indicators can be derived from both observational and modelled time series, to investigate spatial and temporal distribution of intense to extreme conditions.

The recent availability of a new set of simulations, performed as part of the 6th phase of the Coupled Model Intercomparison Project (CMIP6, Eyring et al. 2016), gives the possibility to evaluate the ability of this new generation of Coupled General Circulation Models (CGCMs) to represent intense and extreme precipitation events over the past period and to quantify projected changes in the shape of the tail of the events distribution at the end of the current century.

It is well known that horizontal resolution plays a fundamental role, especially in the representation of precipitation patterns (Haarsma et al. 2016; Roberts et al. 2018), both in terms of averages and extremes, thus we do expect a more accurate representation of precipitation distribution in time and space in the new CMIP6 generation models. Within the CMIP6 bunch of CGCMs, only a few have a horizontal resolution higher than 50 km (up to 25 km of the ones participating to the HighResMIP project; Haarsma et al. 2016), but none of the potential future scenarios to the end of the current century are still available for the community, based on such models. On the other hand, a subset of models having a nominal resolution of 100 km (that corresponds to the highest resolution available within previous CMIP5 scenarios) have already been used to perform the simulation of the worst expected scenarios (the SSP5-RCP8.5) – the one with the highest rate of increase in greenhouse gas concentrations within the new set of scenarios – under the CMIP6 ScenarioMIP protocol (O'Neill et al. 2016). These models are the ones we selected to investigate averages and extremes of precipitation and temperature under a changing climate.

2.2 Modelling Past Extreme Events to Project Future Changes

2.2.1 Climate Models and Simulations

2.2.1.1 Models and Experiment Design As already mentioned, the fully CGCMs employed in this work have been implemented and developed in the framework of the CMIP6. We selected the (five) CGCMs available with a nominal resolution of at least 100 km, providing both historical and SSP5-RCP8.5 future scenarios to the end of the current century. Table 2.1 lists the considered models and summarises their main characteristics.

Three periods are analysed:

- the period 1996–2014 (labelled VALIDATION), corresponding to the common period between the 'historical' CMIP6 simulation and the past period covered by the Global Precipitation Climatology Project (GPCP, Bolvin et al. 2009) used for validation. The same period has been adopted for temperature validation based on JRA-55 re-analysis (Kobayashi et al. 2015).
- the period 1966–2005 (labelled PAST). The choice to end the PAST period in 2005 gives the possibility to compare results from the current assessment to previous CMIP5 (the previous CMIP at the base of the fifth Assessment report of the Intergovernmental Panel on Climate change (IPCC); Taylor et al. 2012) evaluations where the historical period ended in 2005.
- the period 2061–2100 (labelled FUTURE), runs under the high-end CMIP6 'SSP5-RCP8.5' scenario (O'Neill et al. 2016).

In the present analysis, both boreal summer (June to August, JJA) and winter (December to February, DJF) seasons are considered. The historical simulation is performed forcing CMIP6 models with observed concentrations (or emissions, depending on the CGCM implementation) of greenhouse gasses, aerosols, ozone and solar irradiance,

Table 2.1   CMIP6 models involved in the present analysis.

Model acronym	Extended model name	Nominal horizontal resolution	Number of atmospheric grid points [LAT×LON]	Institute (country)
BCC-CSM2-MR	The Beijing Climate Center Climate System Model-2, Medium Resolution	100 km	160 × 320	Beijing Climate Center (China)
CESM2	Community Earth System Model-2	100 km	192 × 288	National Center for Atmospheric Research (USA)
EC-Earth3-Veg	EC-Earth Consortium Model-3-Interactive Vegetation	100 km	256 × 512	European Community Earth-System Model (Europe)
GFDL-ESM4	Geophysical Fluid Dynamics Laboratory-Earth System Model 4	100 km	180 × 288	Geophysical Fluid Dynamics Laboratory (USA)
MRI-ESM2-0	Meteorological Research Institute-Earth System Model 2-0	100 km	160 × 320	Meteorological Research Institute (Japan)

starting from an arbitrary point of a quasi-equilibrium pre-industrial control run. The SSP5-RCP8.5 scenario follows a rising radiative forcing pathway leading to 8.5 W m^{-2} in 2100. The 8.5 W m^{-2} increase of the radiative balance at the top of the atmosphere is referred to the pre-industrial conditions and in the case of the SSP5-RCP8.5 scenario is associated to a CO_2 concentration equivalent to almost three times the pre-industrial 273 ppm value.

Model results are validated based on the last version 1DD-V1.3 of the GPCP (Bolvin et al. 2009) data at 1 degree of horizontal resolution and at the daily timescale available from 1996 to 2018 and JRA-55 re-analysis (Kobayashi et al. 2015). In the present work. GPCP and JRA-55 data (observations hereafter) are limited to 2014 for comparison with the end year of the historical period in CMIP6 protocol. For additional examples of GPCP usage in intense precipitation evaluation, see Shiu et al. (2012), Liu et al. (2009), Scoccimarro et al. (2013), Scoccimarro et al. (2014, 2015, 2016) and Villarini et al. (2014). Similar analysis on extreme events, based on JRA-55 temperature data, can be found in Scoccimarro et al. (2017).

2.2.1.2   Model Ability in Representing the Past Climate Before assessing future climate projection as suggested by the models, it is necessary to evaluate their ability in representing the current climate both in terms of averages and extremes.

Previous assessments (Kharin et al. 2007; O'Gorman and Schneider 2009; Scoccimarro et al. 2013) have shown that climate models provide a realistic representation of present-day heavy precipitation in the extra tropics, but uncertainties in heavy precipitation in the tropics are very large. The 90p and 99p are consistently simulated at mid and high latitudes by CMIP5 models, but they tend to underestimate these indices in the tropics, especially in the northern summer, but also at high latitudes in the

Northern Hemisphere during northern winter and in the Southern Hemisphere during northern summer.

Figures 2.1, 2.2 and 2.3 compare average, intense and extreme precipitation, over the VALIDATION period in CMIP6, in terms of mean, 90th and 99th percentile respectively for both boreal winter (December to February – DJF upper panels) and boreal summer (June to August – JJA lower panels). Model results are shown as ensemble averages and compared to the GPCP observational data set. Figures 2.4, 2.5 and 2.6 show the same comparison but applied to the temperature field, based on JRA-55 re-analysis.

Concerning average precipitation, model results are in agreement with observations with a bias lower than 1 mm d^{-1} over most of the domain in both seasons (Figure 2.1). During DJF (Figure 2.1 upper panels) there is a tendency to overestimate the average precipitation over the Maritime Continent (+3 mm d^{-1}), the southern part of the African continent (+2 mm d^{-1}) and the Andes and eastern Brazil (+5 mm d^{-1}). On the other hand. an underestimation up to –3 mm d^{-1} is expected over Colombia and Venezuela for the same season. During JJA (Figure 2.1, lower panels) the domain affected by a bias higher than 1 mm d^{-1} is even less extended than during DJF, but a

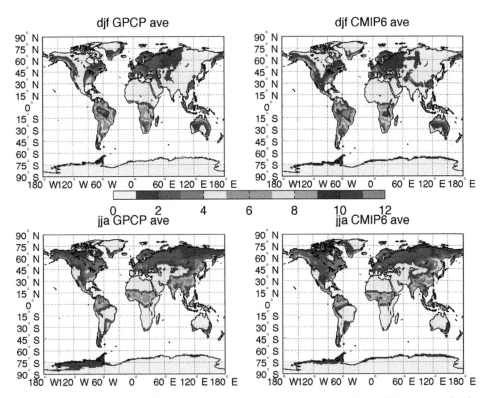

Figure 2.1 Average Precipitation in the present climate. 1996–2014 DJF and JJA averaged values (upper and lower panels respectively) are shown for observations and models (left and right panels respectively). Model results are shown as an ensemble average. Units are [mm d^{-1}].

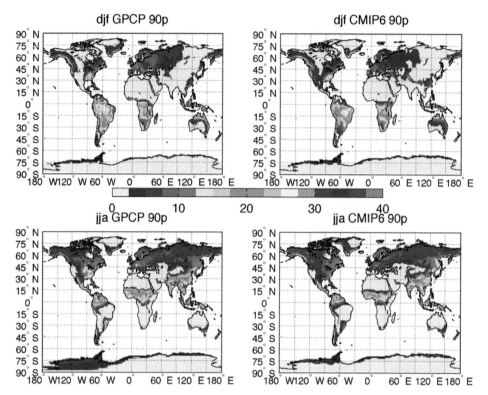

Figure 2.2   Intense Precipitation in the present climate. 1996–2014 DJF and JJA 90th percentile values (upper and lower panels respectively) are shown for observations and models (left and right panels respectively). Model results are shown as ensemble average. Units are [mm/d].

significant underestimation, up to −5 mm d^{-1}, appears over Central America and northern India.

Focusing on intense precipitation (the seasonal 90th percentile), more pronounced biases appear (Figure 2.2). During both seasons, the models tend to underestimate the 90th percentile of precipitation, especially over the tropical belt. An overestimation of intense precipitation appears only along the eastern US coast and Andes during DJF. Moving to the evaluation of models in representing extreme events of precipitation (here represented as the seasonal 99th percentile), huge biases emerge over some regions.

The positive bias of the 99th percentile is mainly confined to the summer season in both hemispheres. There is a general tendency to overestimate extreme precipitation over the wet regions (Figure 2.3) and this is more evident than in previous CMIP5 results (Scoccimarro et al. 2013, 2016) with maximum biases of the order of 40 mm d^{-1}. This highlights a performance degradation in CMIP6 compared to the previous generation of climate models, at least in terms of extreme precipitation representation.

Concerning averaged temperature, CMIP6 models show a bias lower than 1 °C over Europe in both winter and summer seasons (Figure 2.4), better than what we obtained

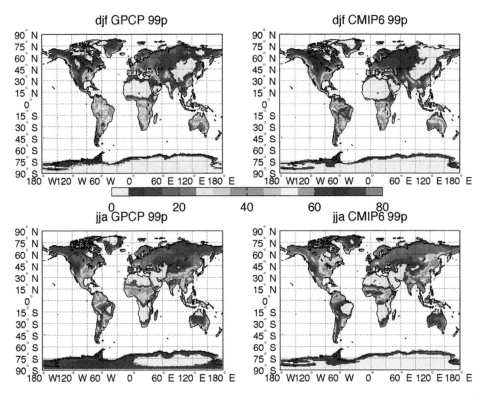

Figure 2.3 Extreme Precipitation in the present climate. 1996–2014 DJF and JJA 99th percentile values (upper and lower panels respectively) are shown for observations and models (left and right panels respectively). Model results are shown as ensemble average. Units are [mm/day].

based on CMIP5 results (Cattieaux et al. 2013). On the other hand, the models tend to underestimate observed averaged temperature over northern Africa (up to –2 °C), northern India (up to –3 °C) and Greenland (up to –5 °C) during DJF. Underestimation of mean surface temperatures over high northern latitudes, particularly in Greenland and Siberia, and for the Tibetan Plateau, was also reported for CMIP3 models (Randall et al. 2007) and for CMIP5 (Silliman et al. 2013). Less pronounced biases are found focusing on JJA when only a small portion of the Asian domain (northern India and north of northern India) shows a positive bias up to 3 °C. The negative bias over the Tibetan plateau, characterising the CMIP5 models (Chen et al. 2017), is confirmed by the CMIP6 model generation.

Focusing on seasonal intense temperature events (90th percentile, Figure 2.5), the models tend to overestimate by up to +5 °C on the northern part of Asia and Canada during DJF and over India, southwestern part of Asia and central and eastern United States during JJA. A positive bias between +2 °C and +3 °C affects the whole of Australia during DJF and is reduced to less than +2 °C during JJA. A negative bias of intense temperature appears only over northern Africa and Saudi Arabia during DJF.

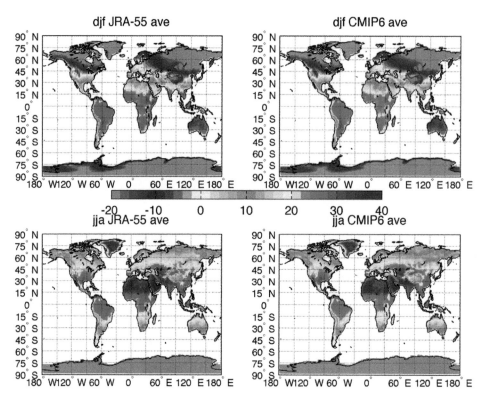

Figure 2.4 Average Temperature in the present climate. 1996–2014 DJF and JJA averaged values (upper and lower panels respectively) are shown for observations and models (left and right panels respectively). Model results are shown as ensemble average. Units are °C.

The aforementioned DJF negative bias persists, also focusing on extreme seasonal temperature (99th percentile, Figure 2.6). Instead the domain affected by the positive bias in the northern part of Asia and Canada is less extended compared to the one associated to intense events in the same season. The positive bias over Australia is confirmed for the 99th percentile, but less pronounced compared to the 90th percentile during both seasons. Overall, the distribution of temperature events is reasonably well represented by CMIP6 models with improved results, compared to previous CMIP efforts, as for instance over Europe.

2.2.2   Observed Changes in Precipitation and Temperature

Before investigating future projections under the CMIP6 SSP5-RCP8.5 scenario for the end of the current century, a literature review is proposed below for the description of observed past tendencies of precipitation and temperature averages and extremes at the global scale.

In the following, we are using probability classes as defined by the IPCC (www.ipcc. ch): each finding is grounded in an evaluation of underlying evidence and agreement. A level of confidence is expressed using five qualifiers: very low, low, medium, high and

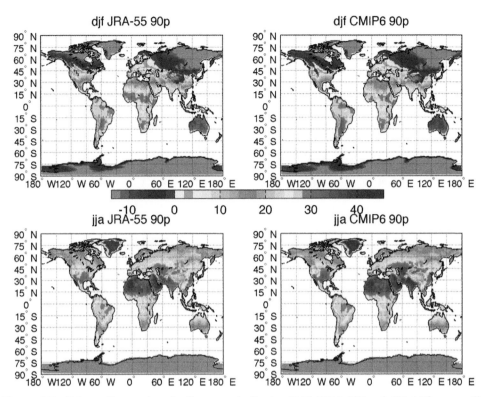

Figure 2.5 Intense Temperature in the present climate. 1996–2014 DJF and JJA 90th percentile values (upper and lower panels respectively) are shown for observations and models (left and right panels respectively). Model results are shown as ensemble average. Units are °C.

very high. The following terms have been used to indicate the assessed likelihood of an outcome or a result: virtually certain 99–100% probability, very likely 90–100%, likely 66–100%, about as likely as not 33–66%, unlikely 0–33%, very unlikely 0–10% and exceptionally unlikely 0–1%.

2.2.2.1 Averaged and Extreme Precipitation Changes in the Past

Both IPCC AR4 (IPCC 2007) and AR5 (IPCC 2014) assessment reports suggest an increase in the number of heavy precipitation events over the second half of the twentieth century over many land regions, even over those where there has been a reduction in total precipitation amounts, consistent with a higher availability of moisture associated with a warming climate and a more pronounced moisture convergence at the surface (Giorgi et al. 2011; Tebaldi et al. 2006). In fact, the capacity of the atmosphere to hold water vapour increases by about 7% per degree Celsius of temperature increase. This increased atmospheric moisture feeds storms on all spatial scales with more water vapour, thus precipitation is expected to increase. In addition, extra precipitation releases more latent heat of condensation in the rising air associated with the convergence of moisture into the storm. This causes enhanced convergence of more moisture, increasing the precipitation rate, thus events with moderate precipitation

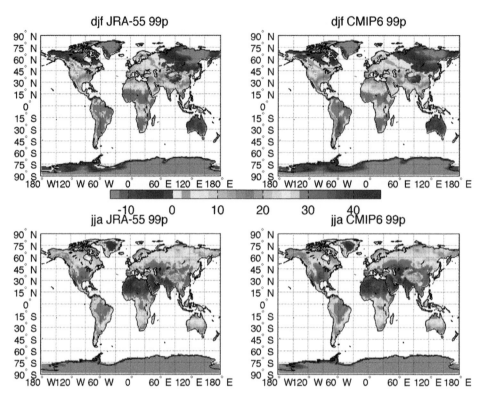

Figure 2.6 Extreme Temperature in the present climate. 1996–2014 DJF and JJA 99th percentile values (upper and lower panels respectively) are shown for observations and models (left and right panels respectively). Model results are shown as ensemble average. Units are °C.

amounts must decrease, while the heavy precipitation events become more frequent. Alexander et al. (2006) highlighted that the observed changes in precipitation extremes are much less spatially coherent and statistically significant compared to observed changes in temperature extremes. There is increasing evidence of a past increase in heavy precipitation but such variations are not statistically significant over some regions. Significant extreme precipitation increase has been found over North America (Kunkel et al. 2008), Canada, the United States and Mexico (Peterson et al. 2008). The evidence indicates a likely increase in observed heavy precipitation in many regions in North America, despite statistically non-significant trends and some decreases in some subregions. This general increase in heavy precipitation follows the increase of the total precipitation in most of the areas. Over Europe, increases in precipitation over land north of 30°N, and decreases over land between 10°S and 30°N, has been registered in the last century. On the other hand, intense precipitation in Europe exhibits complex variability and a lack of a robust spatial pattern, but there are more regions that exhibit a positive rather than a negative trend in heavy precipitation (NAS and NMI 2013).

Positive trends in heavy precipitation have also been registered in Asia (Alexander et al. 2006), Japan (Fujibe et al. 2006), India (Krishnamurthy et al. 2009), China (Zhai et al. 2005) and Australia (Gallant and Karoly 2010). Finally, heavy precipitation tendencies over Africa are difficult to evaluate, due to lack of literature and data. To summarise and to give a measure of the uncertainty associated with observed changes in precipitation extremes, we observe that the number of regions with a likely statistically significant increase in the number of heavy precipitation events (e.g., 95th percentile) is larger than the number of regions with a significant decrease. However, large regional and subregional variations still persist.

2.2.2.2 Averaged and Extreme Temperature Changes in the Past It is well known that the average temperature has increased over the past century, reaching a global average increase of about 1 °C at the end of the last century compared to the pre-industrial value. Such an increase is not constant around the globe, with some region affected by a more pronounced increase such as ice (sea ice and land ice) covered domains and a general tendency to a more pronounced warming over land than over ocean (Hansen et al. 2006).

Regional and global analyses of temperature extremes over land generally suggest a recent increase consistent with a warming climate at the global scale, and this is confirmed in both the 4th and 5th Assessment Reports of the IPCC. Only a few regions are subject to changes in temperature extremes consistent with cooling, most notably for some extremes in central North America, the eastern United States, and also parts of South America. Also, an overall decrease in the number of cold days and nights and an overall increase in the number of warm days and nights in most regions has been found. This applies at the continental scale in North America, Europe and Australia. However, over some sub-regions, the warming trends in temperature extremes are not statistically significant (such as southeastern Europe), and cooling trends are found elsewhere, also in terms of temperature extremes (such as eastern United States and central North America). The Asian region is also affected by trends consistent with warming over most of the continent. On the other hand, similarly to what has already been said about precipitation, Africa is insufficiently well sampled to allow conclusions at the continental scale, despite the few regions with a sufficiently high number of stations exhibited warming in temperature extremes (SREX IPCC 2012). Similar conclusions can be made for South America. Given the measure of the uncertainty associated to past changes in temperature extremes, we have experienced a very likely decrease in number of unusually cold days and nights and a very likely increase in number of unusually warm days and nights at the global scale. There is medium confidence in increase in length or number of warm spells or heatwaves in many (but not all) regions and low or medium confidence in trends in temperature extremes in some subregions, due either to lack of observations or varying signal within subregions (SREX IPCC 2012).

2.2.3 Expected Changes for the End of the Current Century

2.2.3.1 XXI Century Projected Averaged and Extreme Precipitation Changes As already mentioned, the availability of a new set of climate simulations for the twenty-first century has been carried out with state-of-the-art CGCMs produced for phase 6 of the Coupled Model Intercomparison Project (CMIP6). These simulations allow us to investigate future changes in intense precipitation following the most extreme emission pathways (SSP5-RCP8.5), leading to the same radiative balance increase, for the end of the century, of the previous (CMIP5 generation) RCP8.5 scenario. Then we will be able to compare, at least partially, CMIP5 to CMIP6 projections for the end of the current century according to the worst potential scenario.

During the FUTURE period, CMIP5 suggested a slight increase in the amount of rainfall associated with heavy precipitation, over land, in a warmer climate and this is confirmed by CMIP6 results.

Future changes in climatological precipitation patterns (Figure 2.7) are overall coherent with previous findings (Giorgi and Bi 2009; Scoccimarro et al. 2013) obtained using models from the previous phases of the Coupled Model Intercomparison Projects (CMIP3 and CMIP5). During boreal winters a general increase in precipitation over land is found, except for Central and South America. During boreal summers, a general increase of precipitation over land at latitudes higher than 55°N is found and a strong decrease of up to 40% over southern Europe appears (red areas in Figure 2.7, bottom

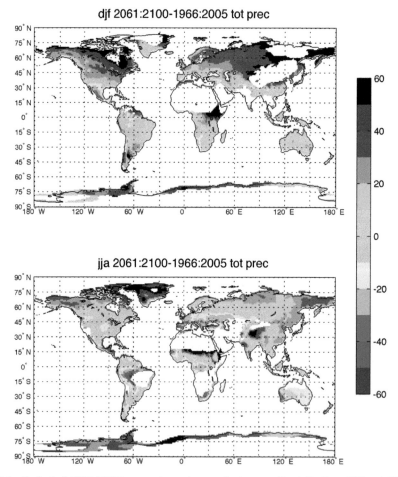

Figure 2.7   Projected SSP5-RCP85 percentage changes in Average Precipitation: DJF and JJA projection of averaged values (upper and lower panels respectively) are shown for the period 2061–2100 with respect to the period 1966–2005. Results are expressed as the ensemble average.

panel). A decrease in the total precipitation also appears over Australia, western North America, Central America, equatorial South America and western equatorial Africa. Precipitation over the Antarctic domain is projected to decrease during both the seasons. The most pronounced increase (reaching 60%) in average precipitation is expected over Siberia, eastern Canada and eastern equatorial Africa during winter and over Greenland and Tibetan Plateau during summer. The most pronounced decrease, reaching –40%, is expected during summer over the Mediterranean regions and Central America. A consistent increase in average precipitation is expected during both winter and summer north of 60 °N.

Future changes in 90p (Figure 2.8) follow the described changes in total precipitation. Patterns are very similar but the magnitude of the projected changes is higher, especially over decreasing precipitation regions such as in the western Mediterranean, where a –60% decrease is reached. Such projected decrease in 90p is more evident in

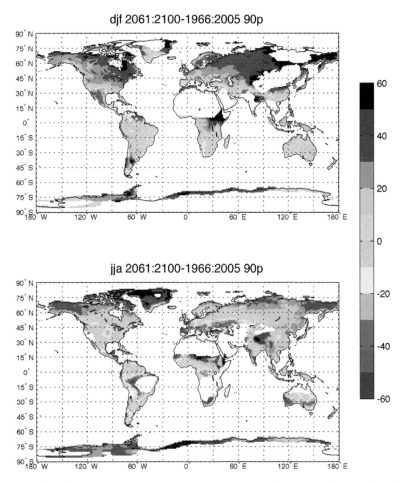

Figure 2.8   Projected SSP5-RCP85 changes in Intense Precipitation: DJF and JJA projection of 90th percentile values (upper and lower panels respectively) are shown for the period 2061–2100 with respect to the period 1966–2005. Results are expressed as the ensemble average.

CMIP6 than what has been previously obtained based on CMIP5 (see Scoccimarro et al. 2013, their Figure 2.4 central panel).

Despite the very similar patterns found in future changes of climatological precipitation and 90p (Figures 2.7 and 2.8 respectively), the 99p changes, pertaining to extreme precipitation events, look different. In the FUTURE period, the 99p increases almost worldwide (Figure 2.9), even over regions where total precipitation and 90p values

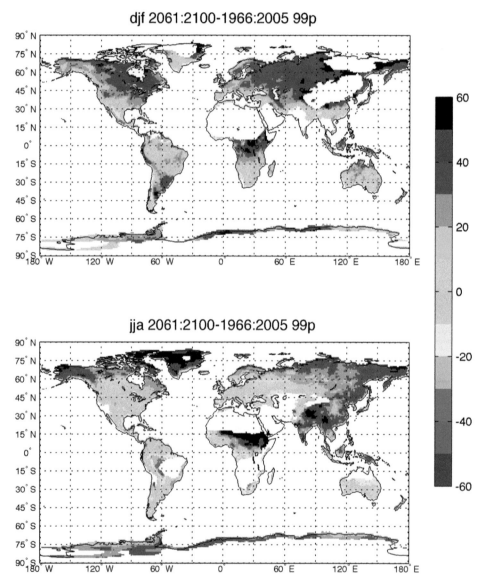

Figure 2.9 Projected SSP5-RCP85 percentage changes in Extreme Precipitation: DJF and JJA projection of 99th percentile values (upper and lower panels respectively) are shown for the period 2061–2100 with respect to the period 1966–2005. Results are expressed as the ensemble average.

show a decrease (red areas in Figures 2.7 and 2.8). This is the case in southern Europe, especially during boreal winter, where the width of the right tail of the distribution increases, even if nearly the entire precipitation distribution becomes dryer (i.e., decreases in total and 90p).

2.2.3.2 XXI Century Projected Averaged and Extreme Temperature Changes Different factors may shape future changes in regional temperature distributions in addition to the direct thermo-dynamical effect expected due to global warming. Temperatures may be affected by potential changes in atmospheric circulation, local mechanisms such as cloud processes or soil moisture feedbacks (Cattiaux et al. 2013). For the evaluation of changes in heat-related disease levels under future projections, not only temperature but also other factors such as environmental relative humidity are important (Scoccimarro et al. 2017). The most important factor affecting heat-related disease is the occurrence of extreme temperature events.

Increases in the frequency and magnitude of warm daily temperature extremes and decreases in cold extremes will occur throughout the twenty-first century at the global scale (SREX IPCC 2012). It is very likely that the length, frequency and/or intensity of warm spells or heatwaves will increase over most land areas. In terms of absolute values, 20-year extreme annual daily maximum temperature (i.e., return value) will likely increase by 3 °C by the mid-twenty-first century and by about 5 °C by the late twenty-first century, based on the worst emission scenario considered. Anyway, regional changes in temperature extremes will often differ from the mean global temperature change.

In the following paragraph, we investigate average and extreme temperature projections based on the new CMIP6 SSP5-RCP8.5 scenario and also discuss regional changes.

An average temperature increase higher than 2 °C is expected over land during both winter and summer, with the most pronounced increase in the Northern Hemisphere, north of 60 °N during DJF. The polar amplification is expected to continue, and projected changes are consistent with losses of sea ice and snow retreat. The lower the albedo, the more a surface absorbs heat from sunlight rather than reflecting it back into space. This winter temperature increase is expected to be even higher than 12 degrees over certain terrain regions of Canada and Siberia (Figure 2.10, upper panel), while during summer the warming of these regions (north of 60°N) of land follows more or less what is expected for the rest of the extratropical Northern Hemisphere. The CMIP6 averaged temperature projections also show that the thermal contrast between the Northern and Southern Hemispheres, an index of climate change, is projected to increase at the end of the century. During both winter and summer seasons, a more evident warming of the Northern Hemisphere is expected, even considering land only (in addition to the expected stronger hemispheric increase due to the higher fraction of land compared to the Southern Hemisphere). This is expected to strongly impact the general circulation by weakening the NH Hadley cell, strengthening the SH Hadley cell, and shifting tropical precipitation north (Friedman et al. 2013).

The projection of intense events (90p) is similar to what we expect for averaged temperature, but with a less extended domain affected by more than a 11 °C increase in the Northern Hemisphere during DJF and a more pronounced increase over the central United States and Europe, especially during JJA, exceeding 8 °C (Figure 2.11). This

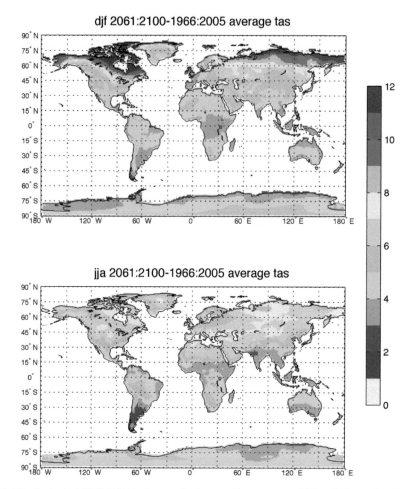

Figure 2.10   Projected SSP5-RCP85 changes in Average Temperature: DJF and JJA projection of averaged values (upper and lower panels respectively) are shown for the period 2061–2100 with respect to
the period 1966–2005. Results are expressed as the ensemble average. Units are in °C.

tendency is even more evident focusing on extreme events (99p) of temperature. The
domain affected by the aforementioned huge increase (higher than 11 °C) in the north-
ward part of the Northern Hemisphere during DJF is reduced by more than 80% compared to average temperature projections (Figure 2.12). In addition, the 99p increase
over the central United States and Europe during JJA is projected to reach 9 °C. An
interesting metric defined to investigate the width of the right tail of an event distribution is the 99–90p (Scoccimarro et al. 2013), defined as the distance between the 99th
and the 90th percentile of the event distribution. Figure 2.13 shows the projections of
the 99–90p for the end of the century. Interesting, we do expect a reduction of the width
of the right part of the temperature events distribution over most of the regions at latitudes higher than 50°, especially during DJF, thus over regions covered by ice or snow
in the current climate. On the other hand, most of the land domain at latitudes lower
than 50° is expected to be affected by an enlargement of the width of the right tail of

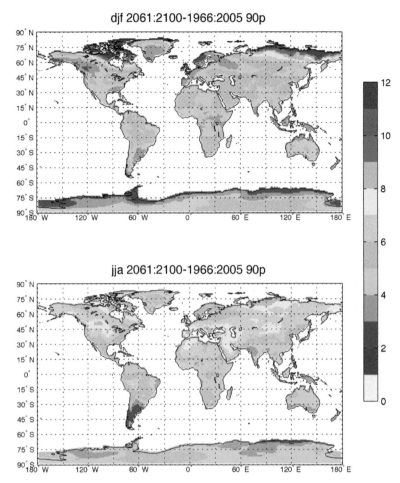

Figure 2.11 Projected SSP5-RCP85 changes in Intense Temperature: DJF and JJA projection of 90th percentile values (upper and lower panels respectively) are shown for the period 2061–2100 with respect to the period 1966–2005. Results are expressed as the ensemble average. Units are in °C.

the temperature distribution with maximum stretching expected over central South America, western Asia and southern parts of Africa during DJF. During JJA, the maximum projected stretching of the tail of the distribution is expected over the Arctic Canadian archipelago (even higher than 2 °C) and Indochina. Anyway, during both seasons, the whole tropical domain is affected by a stretching of the tail of the temperature event distribution higher than 0.5 °C. Such an increase represents a non-negligible stretching of the distribution tail since the average 99–90p diagnostic ranges from 2 to 3 °C (not shown) over the tropics in both seasons.

This is highlighting a projected change of the shape of the daily temperature event distribution (a stretched right tail of the distribution), particularly evident over the tropical regions, for the end of the current century, following the worst warming scenario considered by CMIP6.

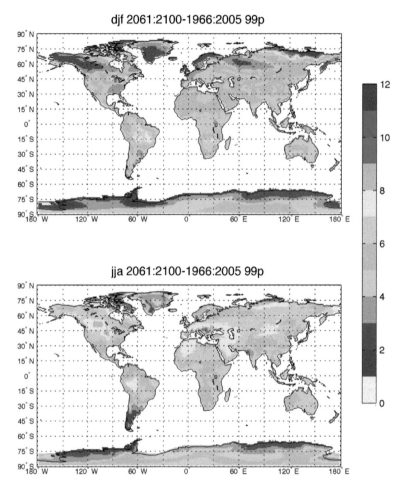

Figure 2.12 Projected SSP5-RCP85 changes in Extreme Temperature: DJF and JJA projection of 99th percentile values (upper and lower panels respectively) are shown for the period 2061–2100 with respect to the period 1966–2005. Results are expressed as the ensemble average. Units are in °C.

2.3   Conclusions

Extensive work has been done in the past to evaluate projected changes of precipitation and temperature distributions for the end of the current century. Our results indicate that CMIP6 has similar difficulties as the CMIP5 models to represent precipitation, in particular regarding the representation of extreme precipitation. CMIP6 models tend to overestimate extreme precipitation over wet regions.

The projected patterns of precipitation are very similar between CMIP5 and CMIP6, with some differences in terms of intensity, but with a consistent spatial distribution of changes between the two relative scenarios chosen to represent the worst emission scenario (RCP8.5 and SSP5-RCP8.5 respectively).

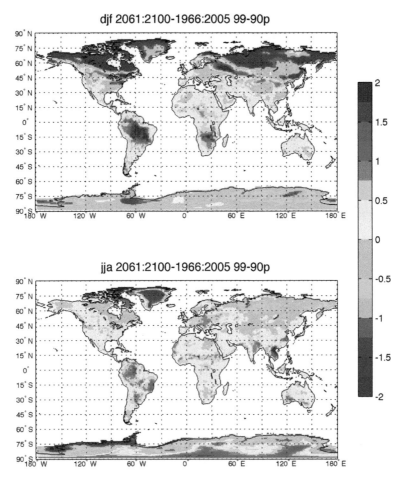

Figure 2.13 Projected SSP5-RCP85 changes in the 99-90p diagnostic: DJF and JJA projection of the difference between 99th and 90th percentile values (upper and lower panels respectively) are shown for the period 2061–2100 with respect to the period 1966–2005. This diagnostic is representative of the width of the right tail of the event distribution. Results are expressed as the ensemble average. Units are [oC].

In general, current models suggest an increase of extreme precipitation almost worldwide with few exceptions, while in terms of averages we also expect regions affected by a substantial drying such as southern Europe, central United States and South America, mainly during the boreal summer. Temperature projections are consistent with CMIP5 results, with more pronounced increases over some regions. The changes of the probability distribution of daily temperature events in the tropical regions show that the stretching of the tail of the temperature event distribution reaches 20–30% beyond the distribution tail representative of the historical period, indicating an increase in large events.

This last finding has important implications in terms of impacts, especially concerning human health, since a more pronounced increase of extremely high temperatures might affect a population more than what we can expect from the increase of the average temperature. The implications of such results are large, also because they mainly affect regions already exposed to conditions of extreme heat at present. This will impact heat-related mortality, typically associated to pre-existing chronic conditions (Fouillet et al. 2006; Haines et al. 2006), but also will increase vulnerability to heat-related illness, including people working outdoors or in non-cooled environments (Hanna et al. 2011; Yin and Wang 2017).

Losses from weather and climate disasters are rising (Mechler et al. 2019), with the increase largely due to increased exposure to extreme events. Past studies have shown (Scoccimarro et al. 2013, 2015; SREX IPCC 2012) – and this is also confirmed by the CMIP6 projections discussed in the present chapter – that a warming world will likely lead to more extreme natural environments and human activity is likely to be driving most of the trends. Adaptation and disaster risk management can enhance resilience in a warmer climate and differences in vulnerability and exposure must be considered in the design of such initiatives (see Chapters 5 and 6). While local action can help to reduce local risks, more transformative changes to governance and technological systems will also be required.

References

Alexander, L.V., Zhang, X., Peterson, T.C. et al.(2006). Global observed changes in daily climate extremes of temperature and precipitation. *Journal of Geophysical Research – Atmospheres* 111: D05109.

Ballester, J., Giorgi, F., and Rodo, X. (2010). Changes in European temperature extremes can be predicted from changes in PDF central statistics. *Climatic Change* 98 (1–2): 277–284.

Bell, J.E., Brown, C.L., Conlon, K. et al. (2018). Changes in extreme events and the potential impacts on human health. *Journal of the Air & Waste Management Association* 68 (4): 265–287. doi:10.1080/10962247.2017.1401017.

Bolvin, D.T., Adler, R.F., Huffman, G.J. et al. etal.(2009). Comparison of GPCP monthly and daily precipitation estimates with high-latitude gauge observations. *Journal of Applied Meteorology and Climatology* 48: 1843–1857.

Brown, S.J., Caesar, J., and Ferro, C.A.T. (2008). Global changes in extreme daily temperature since 1950. *Journal of Geophysical Research – Atmospheres* 113: D05115.

Cattiaux, J., Douville, H., and Peings, Y. (2013). European temperatures in CMIP5: Origins of present-day biases and future uncertainties. *Climate Dynamics* 41 (11–12): 2889–2907. doi:10.1007/s00382-013-1731-y.

Chen, X.L., Liu, Y.M., and Wu, G.X. (2017). Understanding the surface temperature cold bias in CMIP5 AGCMs over the Tibetan Plateau. *Advances in Atmospheric Sciences* 34: 1447–1460. doi:10.1007/s00376-017-6326-9.

Della-Marta, P.M., Haylock, M.R., Luterbacher, J. et al. (2007). Doubled length of western European summer heat waves since 1880. *Journal of Geophysical Research – Atmospheres* 112: D15103.

EEA, 2019: Economic losses from climate-related extremes in Europe. https://www.eea.europa.eu/data-and-maps/indicators/direct-losses-from-weather-disasters-3/assessment-2.

Eyring, V., Bony, S., Meehl, G.A. et al. (2016). Overview of the coupled model intercomparison project phase 6 (CMIP6) experimental de 5 sign and organisation. *Geoscientific Model Development Discussions* 8: 10539–10583. doi:10.5194/gmdd-8-10539-2015.

Fouillet, A., Rey, G., Laurent, F. et al. (2006). Excess mortality related to the August 2003 heat wave in France. *International Archives of Occupational and Environmental Health* 80 (1): 16–24. 2006.

Friedman, A.R., Hwang, Y., Chiang, J.C. et al. (2013). Interhemispheric temperature asymmetry over the twentieth century and in future projections. *Journal of Climate* 26: 5419–5433. https://doi.org/10.1175/JCLI-D-12-00525.1.

Fujibe, F., Yamazaki, N., and Kobayashi, K. (2006). Long-term changes of heavy precipitation and dry weather in Japan (1901–2004). *Journal of the Meteorological Society of Japan* 84 (6): 1033–1046.

Gallant, A.J.E. and Karoly, D.J. (2010). A combined climate extremes index for the Australian region. *Journal of Climate* 23 (23): 6153–6165.

Giorgi, F. and Bi, X. (2009). Time of emergence (TOE) of GHG forced precipitation change hotspots. *Geophysical Research Letters* 36: L06709. doi:10.1029/2009GL037593.

Giorgi, F., Im, E.-S., Coppola, E., Diffenbaugh, N.S. et al. (2011). Higher hydroclimatic intensity with global warming. *Journal of Climate* 24: 5309–5324. doi:10.1175/2011JCLI3979.1.

Griffiths, G.M., Chambers, L.E., Haylock, M.R. et al. (2005). Change in mean temperature as a predictor of extreme temperature change in the Asia-Pacific region. *International Journal of Climatology* 25 (10): 1301–1330.

Haarsma, R.J., Roberts, M., Vidale, P.L. et al. (2016). High resolution model intercomparison project (HighResMIP). *Geoscientific Model Development.* doi:10.5194/gmd-2016-66.

Haines, A., Kovats, R.S., Campbell-Lendrum, D. et al. (2006). Climate change and human health: Impacts. *Vulnerability, and Mitigation Lancet* 367 (9528): 2101–2109.

Hanna, E.G., Kjellstrom, T., Bennett, C. et al. (2011). Climate change and rising heat: Population health implications for working people in Australia. *Asia-Pacific Journal of Public Health* 23 (2): 14S–26S. 2011.

Hansen, J., Sato, M., Ruedy, R. et al. (2006). Global temperature change. *Proceedings of the National Academy of Sciences* 103: 14288–14293. doi:10.1073/pnas.0606291103.

IPCC (2007). *Climate Change 2007: Synthesis Report. Contribution of Working Groups I, II and III to the Fourth Assessment Report of the Intergovernmental Panel on Climate Change.* Core Writing Team (ed. R.K. Pachauri and A. Reisinger), 104. Geneva, Switzerland: IPCC.

IPCC (2014). *Climate Change 2014: Synthesis Report. Contribution of Working Groups I, II and III to the Fifth Assessment Report of the Intergovernmental Panel on Climate Change.* Core Writing Team (ed. R.K. Pachauri and L.A. Meyer), 151. Geneva, Switzerland: IPCC.

Kharin, V.V., Zwiers, F.W., Zhang, X. et al. (2007). Changes in temperature and precipitation extremes in the IPCC ensemble of global coupled model simulations. *Journal of Climate* 20: 1419–1444.

Kobayashi, S., Ota, Y., Harada, Y. et al. (2015). The JRA-55 Re-analysis: General specifications and basic characteristics. *Journal of the Meteorological Society of Japan* 93: 5–48. doi:10.2151/jmsj.2015-001.

Krishnamurthy, C.K.B., Lall, U., and Kwon, H.H. (2009). Changing frequency and intensity of rainfall extremes over India from 1951 to 2003. *Journal of Climate* 22 (18): 4737–4746.

Kunkel, K.E., Bromirski, P.D., Brooks, H.E. et al. (2008). Observed changes in weather and climate extremes. In: *Weather and Climate Extremes in a Changing Climate. Regions of Focus: North America, Hawaii, Caribbean, and U.S. Pacific Islands* (ed. T.R. Karl, G.A. Meehl, D.M. Christopher, et al.), 222. Washington, DC: A Report by the U.S. Climate Change Science Program and the Subcommittee on Global Change Research.

Liu, S.C., Fu, C., Shiu, C.-J. et al. (2009). Temperature dependence of global precipitation extremes. *Geophysical Research Letters* 36: L17702. doi:10.1029/2009GL040218.

Mechler, R., Bouwer, L., Schinko, T. et al. (2019). Loss and damage from climate change: Concepts, methods and policy options. Springer. https://link.springer.com/book/10.1007/978-3-319-72026-5.

NAS and NMI (2013) Extreme weather events in Europe: Preparing for climate change adaptation. Oslo: Norwegian Academy of Science and Letters and the Norwegian Meteorological Institute. http://www.dnva.no/binl/download.php?tid=58783. ISBN (electronic) 978-82-7144-101-2.

NOAA (2019). Smith, A., NOAA technical report no. 2021Q2. National Centers for Environmental Information (NCEI). U.S. Billion-Dollar Weather and Climate Disasters. https://www.ncdc.noaa.gov/billions.

O'Gorman, P.A. and Schneider, T. (2009). The physical basis for increases in precipitation extremes in simulations of 21st century climate change. *Proceedings of the National Academy of Sciences of the United States of America* 106: 14773–14777.

O'Neill, B.C., Tebaldi, C., Van Vuuren, D.P. et al. (2016). The Scenario Model Intercomparison Project (ScenarioMIP) for CMIP6, Geosci. *Model Development* 9: 3461–3482. doi:10.5194/gmd-9-3461-2016.

Orlowsky, B. and Seneviratne, S.I. (2011). Global changes in extremes events: Regional and seasonal dimension. *Climatic Change*. doi:10.1007/s10584-011-0122-9.

Parry, M.L., Canziani, O.F., Palutikof, J.P. et al. (2007). Climate change 2007. In: *Impacts, Adaptation and Vulnerability* (ed. S. Solomon). Cambridge: Cambridge University Press.

Peterson, T.C., Anderson, D.M., Cohen, S.J. et al. (2008). Weather and climate extremes in a changing climate. Regions of focus: North America, Hawaii, Caribbean, and U.S. Pacific Islands. In: *Synthesis and Assessment Product 3.3* (ed. T.R. Kar), 11–34. Washington, DC: US Climate Change Science Program.

Randall, D., Wood, R.A., Bony, S. et al. (2007). Climate Models and Their Evaluation. In: *Climate Change 2007: The Physical Science Basis. Contribution of Working Group I to the Fourth Assessment Report of the Intergovernmental Panel on Climate Change.* IPCC WG1 Fourth Assessment Report. Cambridge, UK and New York, USA: Cambridge University Press.

Roberts, M., Vidale, P., Senior, C. et al. (2018). The benefits of global high resolution for climate simulation: Process-understanding and the enabling of stakeholder decisions at the regional scale. *Bulletin of the American Meteorological Society.* doi:10.1175/BAMS-D-15-00320.1.

Russo, S., Dosio, A., Graversen, R.G. et al. (2014). Magnitude of extreme heat waves in present climate and their projection in a warming world. *Journal of Geophysical Research Atmospheres* 119: 12500–12512. doi:10.1002/2014JD022098.

Scoccimarro, E., Bellucci, A., Zampieri, M. et al. (2013). Heavy precipitation events in a warmer climate: Results from CMIP5 models. *Journal of Climate* 26: 7902–7911. doi:10.1175/JCLI-D-12-00850.1.

Scoccimarro, E., Bellucci, A., Zampieri, M. et al. (2016). Heavy precipitation events over the Euro-Mediterranean region in a warmer climate: Results from CMIP5 models. *Regional Environmental Change* doi:10.1007/s10113-014-0712-y.

Scoccimarro, E., Fogli, P.G., and Gualdi, S. (2017). The role of humidity in determining perceived temperature extremes scenarios in Europe. *Environmental Research Letters.* doi:10.1088/1748-9326/aa8cdd.

Scoccimarro, E., Gualdi, S., Villarini, G. et al. (2014). Intense precipitation events associated with landfalling tropical cyclones in response to a warmer climate and increased CO_2. *Journal of Climate.* doi:10.1175/JCLI-D-14-00065.1.

Scoccimarro, E., Villarini, G., Vichi, M. et al. (2015). Projected changes in intense precipitation over Europe at the daily and sub-daily timescales. *Journal of Climate.* doi:10.1175/JCLI-D-14-00779.1.

Seneviratne, S.I., Nicholls, N., Easterling, D. et al. (2012). Changes in climate extremes and their impacts on the natural physical environment. In: *Managing the Risks of Extreme Events and Disasters to Advance Climate Change Adaptation* (ed. C.B. Field, V. Barros, T.F. et al.), 109–230. A Special Report of Working Groups I and II of the Intergovernmental Panel on Climate Change (IPCC). Cambridge, UK, and New York, USA: Cambridge University Press.

Shiu, C.-J., Liu, S.C., Fu, C. et al. (2012). How much do precipitation extremes change in a warming climate? *Geophysical Research Letters* 39: L17707. doi:10.1029/2012GL052762.

Sillmann, J., Kharin, V.V., Zhang, X. et al. (2013). Climate extremes indices in the CMIP5 multimodel ensemble: Part 1. Model evaluation in the present climate. *Journal of Geophysical Research Atmospheres* 118: 1716–1733. doi:10.1002/jgrd.50203.

Sillmann, J. and Roeckner, E. Climatic Change (2008). Indices for extreme events in projections of anthropogenic. *Climate Change* 86: 83. https://doi.org/10.1007/s10584-007-9308-6.

Smith, A. and Matthews, J. (2015). Quantifying uncertainty and variable sensitivity within the U.S. billion-dollar weather and climate disaster cost estimates. *Natural Hazards*. doi:10.1007/s11069-015-1678-x.

SREX IPCC (2012). *Managing the Risks of Extreme Events and Disasters to Advance Climate Change Adaptation* (Ed. C.B. Field, V. Barros, T.F. Stocker, et al.), 582. A Special Report of Working Groups I and II of the Intergovernmental Panel on Climate Change. Cambridge, UK, and New York, US: Cambridge University Press.

Taylor, K.E., Stouffer, R.J., and Meehl, G.A. (2012). An overview of CMIP5 and the experiment design. *Bulletin of the American Meteorological Society* 93: 485–498.

Tebaldi, C., Hayhoe, K., Arblaster, M.J. et al. (2006). Going to the extremes: An intercomparison of modelsimulated historical and future changes in extreme events. *Climatic Change* 79: 185–211. doi:10.1007/s10584-006-9051-4.

Villarini, G., Lavers, D.A., Scoccimarro, E. et al. (2014). Sensitivity of tropical cyclone rainfall to idealized global scale forcings. *Journal of Climate*. doi:10.1175/JCLI-D-13-00780.1.

Yin, Q. and Wang, J.F. (2017). The association between consecutive days' heat wave and cardiovascular disease mortality in Beijing, China. *BMC Public Health* 17.

Zampieri, M., Russo, S., Di Sabatino, S. et al. (2016). Global estimation of heat wave magnitudes from 1901 to 2010 and possible implications for the river discharge of the Alps. *Science of the Total Environment*. doi:10.1016/j.scitotenv.2016.07.008.

Zhai, P.M., Zhang, X., Wan, H. et al. (2005). Trends in total precipitation and frequency of daily precipitation extremes over China. *Journal of Climate* 18 (7): 1096–1108.

3
Climate Change and Health

Alistair Woodward
University of Auckland, New Zealand

3.1 Introduction

This chapter aims to provide a broad outline of what is presently known about climate change and human health. The literature on this topic is growing rapidly, in line with the exponential rise in the number of publications on climate change more broadly (Callaghan et al. 2020). The Assessment Reports of the Intergovernmental Panel on Climate Change (IPCC) provide a scientific benchmark, and are published every 6–7 years. This chapter will summarise the findings of the most recent (at the time of writing) assessment report (AR5), published in 2014 (Woodward et al. 2014), and point to significant studies that have been published since. This is a selection, not a systematic sampling, with a particular focus on recent research that is relevant to the subject of this book, hydrometeorological extreme events (HEEs).

3.2 The IPCC 5th Assessment Report

The reports of the IPCC draw on the work of hundreds of scientists and are intensively (some might say exhaustively) reviewed (Woodward et al. 2014). In AR5, human health was included as a chapter in its own right in the report of Working Group 2, which was concerned with impacts, vulnerabilities and adaptation. Climate sciences were the remit of Working Group 1, and mitigation was the focus of Working Group 3.

The executive summary of the human health chapter is contained in Box 3.1.

When reading the IPCC summary, it is important to bear in mind that most of the literature cited here draws on research about the relation between weather and health. Much less is written about impacts of climate ('average weather') and health studies of climate change, which require observations extending over decades or more, are rarer still. Studies of heatwaves, for instance, commonly relate daily temperatures to deaths and illness occurring shortly afterwards. It is inferred that climate change will lead to more frequent heat events, and hence cause health losses, but it is not possible to observe these effects directly.

Hydrometeorological Extreme Events and Public Health, First Edition. Edited by Franziska Matthies-Wiesler and Philippe Quevauviller.
© 2022 John Wiley & Sons Ltd. Published 2022 by John Wiley & Sons Ltd.

Box 3.1   Executive Summary, Chapter 11. Human Health: Impacts, Adaptation and Co-Benefits. From the Report of Working Group 2, IPCC 5th Assessment Report. 2014 (Woodward et al. 2014)

The health of human populations is sensitive to shifts in weather patterns and other aspects of climate change [*very high confidence*]. These effects occur directly, due to changes in temperature and precipitation and occurrence of heatwaves, floods, droughts, and fires. Indirectly, health may be damaged by ecological disruptions brought on by climate change (crop failures, shifting patterns of disease vectors), or social responses to climate change (such as displacement of populations following prolonged drought). Variability in temperatures is a risk factor in its own right, over and above the influence of average temperatures on heat-related deaths. Biological and social adaptation is more difficult in a highly variable climate than one that is more stable.

Until mid-century, climate change will act mainly by exacerbating health problems that already exist [*very high confidence*]. New conditions may emerge under climate change [*low confidence*], and existing diseases (e.g., food-borne infections) may extend their range into areas that are presently unaffected [*high confidence*]. But the largest risks will apply in populations that are currently most affected by climate-related diseases. Thus, for example, it is expected that health losses due to climate change-induced undernutrition will occur mainly in areas that are already food-insecure.

In recent decades, climate change has contributed to levels of ill health [*likely*] though the present worldwide burden of ill health from climate change is relatively small compared with other stressors on health and is not well quantified. Rising temperatures have increased the risk of heat-related death and illness [*likely*]. Local changes in temperature and rainfall have altered distribution of some water-borne illnesses and disease vectors, and reduced food production for some vulnerable populations [*medium confidence*].

If climate change continues as projected across the RCP scenarios until mid-century, the major increases of ill health compared to no climate change will occur through:

- greater risk of injury, disease, and death due to more intense heatwaves and fires [very high confidence];
- increased risk of undernutrition resulting from diminished food production in poor regions [high confidence];
- consequences for health of lost work capacity and reduced labor productivity in vulnerable populations [high confidence];
- increased risks of food- and water-borne diseases [very high confidence] and vector-borne diseases [medium confidence];
- modest improvements in cold-related mortality and morbidity in some areas due to fewer cold extremes [low confidence], geographical shifts in food production, and reduced capacity of disease-carrying vectors due to excedance of thermal thresholds [medium confidence]. These positive effects will be outweighed, worldwide, by the magnitude and severity of the negative effects of climate change [high confidence].

Impacts on health will be reduced, but not eliminated, in populations that benefit from rapid social and economic development [*high confidence*], particularly among the poorest and least healthy groups [*very high confidence*]. Climate change is an impediment to continued health improvements in many parts of the world. If economic growth does not benefit the poor, the health effects of climate change will be exacerbated.

In addition to their implications for climate change, essentially all the important Climate Altering Pollutants (CAPs) other than CO_2 have near-term health implications [*very high confidence*]. In 2010, more than 7% of the global burden of disease was due to inhalation of these air pollutants [*high confidence*].

Some parts of the world already exceed the international standard for safe work activity during the hottest months of the year. The capacity of the human body to thermoregulate may be exceeded on a regular basis, particularly during manual labour, in parts of the world during this century. In the highest Representative Concentration Pathway, RCP8.5, by 2100 some of the world's land area will be experiencing 4–7 degree higher temperatures due to anthropogenic climate change [WG1, Figure SPM.7]. If this occurs, the combination of high temperatures and high humidity will compromise normal human activities, including growing food and other work outdoors, raising doubt about the habitability of some areas, for parts of the year [*high confidence*].

The most effective adaptation measures for health in the near-term are programs that implement basic public health measures such as provision of clean water and sanitation, secure essential healthcare including vaccination and child health services, increase capacity for disaster preparedness and response, and alleviate poverty [*very high confidence*]. In addition, there has been progress since AR4 in targeted and climate-specific measures to protect health, including enhanced surveillance and early warning systems.

There are opportunities to achieve co-benefits from actions that reduce emissions of CAPs and at the same time improve health. Among others, these include:

- reducing local emissions of health-damaging and climate-altering air pollutants from energy systems, through improved energy efficiency, and a shift to cleaner energy sources [very high confidence];
- providing universal access to reproductive health services (including modern family planning) to improve child and maternal health through birth spacing and reduce population growth, energy use, and consequent CAP emissions over time [medium confidence];
- shifting consumption away from animal products, especially from ruminant sources, in high-meat-consumption societies towards less CAP-intensive healthy diets [medium confidence];
- designing transport systems that promote active travel and reduce use of motorized vehicles, leading to lower emissions of CAPs and better health through improved air quality and greater physical activity [high confidence].

There are important research gaps regarding the health consequences of climate change and co-benefits actions, particularly in low-income countries. There are now opportunities to use existing longitudinal data on population health to investigate how climate change affects the most vulnerable populations. Another gap concerns the scientific evaluation of the health implications of adaptation measures at community and national levels. A further challenge is to improve understanding of the extent to which taking health co-benefits into account can offset the costs of GHG mitigation strategies.

Here are two examples of recent papers that have attempted to push back the time horizon in climate change and health research. Lai et al. (2020) examined the relation in New Zealand between heavy rainfall and admission of children to hospital with diarrheal disease. They found the risk of such admissions increased by about 70% 1–2 days after daily rainfall greater than the 95th percentile for that location. The authors then investigated whether the effect of heavy precipitation was modified by long-term rainfall (as measured over the previous 6 years), and found the increase in hospitalisations was strongest in locations that were particularly dry (long-term rainfall in the lowest quintile nationally) or particularly wet (in the highest quintile). The findings are consistent with those of a previous study in the US, finding the risk of salmonellosis following extreme rainfall events was modified by antecedent rainfall conditions (Lee et al. 2019). The essential point here is that there are two exposures that appear to influence rates of water-borne disease: daily rainfall (a feature of weather) and long-term conditions that help define the climate at a given location. Both may be disturbed by climate change. More intense precipitation is projected for many places, where the variability of rainfall may also increase. In New Zealand, for instance, annual rainfall is projected to increase substantially in the west and south of the country, and to decline in eastern regions.

Bennett et al. (2013) examined the seasonal distribution of deaths in Australia from 1968 to 2007, and found the ratio of summer to winter deaths increased steadily in association with a multi-decadal warming trend. This is one of the few instances of work that directly relates climate change to a human health outcome. The implication is that steadily rising temperatures are changing the balance of causes of mortality, increasing deaths that tend to occur in summer and reducing winter-related deaths. So far, it appears to be the latter that is most important. Extension of this analysis to 2020 found the trend continues, with the steepest rise in the proportion of deaths occurring in summer occurring in the hottest years (Hanigan et al. 2021).

The IPCC report concluded that until the mid-century climate change will act mainly by exacerbating health problems that already exist (Woodward et al. 2014). But the report might have signalled more strongly that while entirely new conditions are unlikely, this does not mean that present-day public health arrangements will be sufficient in the future. A good example are the changes required in Nepal to manage dengue fever. The mosquitoes that spread the dengue virus are sensitive to acute changes in temperature and rainfall, as shown in 2019 by a severe outbreak of dengue (more than 10,000 cases) following the heaviest rainfall in Nepal for more than a decade. But the epidemiology of the disease is affected also by long-term climate conditions, and rising temperatures (more marked at higher altitudes) are one of the reasons for the spread of dengue into mountain districts (Pandey et al. 2019). It has been necessary for authorities in Nepal to alter training programmes for environmental health staff. In the past, only workers in lowland areas were concerned with dengue prevention – now public health workers in the mountains must also learn about mosquito control.

As noted in Box 3.1, climate forcing agents may affect human health in other ways than through climate change itself. Many of the greenhouse pollutants released by the burning of fossil fuels have direct, deleterious effects on health. Some of the most severe and best-known episodes of severe air pollution (the great London Smog of 1952, for example) resulted from intense coal burning, and coal was also for almost a hundred years the fuel mostly responsible for rising levels of atmospheric CO_2 (Smith et al. 2013). Air pollution due to burning biomass fuels indoors causes roughly 4 million premature deaths a year worldwide; household fires also release about a gigaton of CO_2e per year (about the same greenhouse load as that attributed to aviation) (Pachauri et al. 2013).

3.3   What Is New Since AR5?

The authors of the Fifth Assessment Report divided the effects of climate change on human health into those resulting from direct impacts, those caused indirectly through ecosystem changes, and effects linked to social disruption due to climate change. In each of these categories there is growing evidence of present-day impacts.

Rising temperatures and changes in precipitation lead directly to increased potential for fires (Di Virgilio et al. 2019). The Australian Black Summer fires of 2019/20 burnt 20 million hectares of forest and destroyed about 6000 structures, and it is estimated the fires killed an astonishing 1.5 billion animals (van Oldenborgh, 2021). What was unprecedented was the location of the fires (in the heavily populated southeast of the country), what was burnt (more than 20% of temperate broadleaf forests in the country, a greater loss in a single season than any observed before, globally), the early onset and duration of the fires (beginning in September, continuing into February) and the scale of the impacts – the Gospers Mountain fire near Sydney was the largest fire recorded in Australia (Nolan et al. 2020). About 30 human lives were lost in the acute phase of the fires and millions of people were exposed to smoke which blanketed major cities for weeks and drove up fine particle levels well above WHO guidelines (Vardoulakis et al. 2020). A survey of a representative sample of the national Australian population in January 2020 found that more than half the respondents had been directly affected by the fires, and a quarter reported adverse health effects from the smoke, including respiratory conditions and difficulty in breathing (Vardoulakis et al. 2020). On the basis of estimated personal exposures, bushfire smoke caused in excess of 417 deaths and 1124 hospital admissions for cardiovascular conditions (Borchers Arriagada et al. 2020). One-third of the deaths were attributed to delayed or disrupted healthcare, as many residents were unable to access essential medicines or medical devices, and clinics were frequently closed.

The influence of long-term climate change on elements of fire risk is evident – 2019 was the hottest and driest year recorded in Australia (Figure 3.1), the fires followed long-term drought in eastern Australia, and moisture levels in forests in New South Wales and Victoria, the two states most affected by the fires, were at an all-time low. A

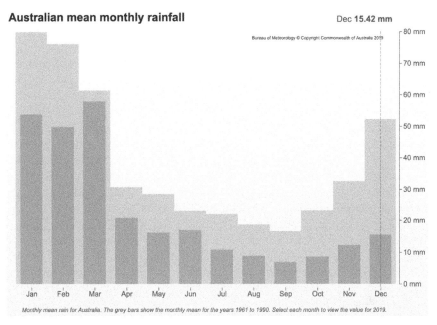

Figure 3.1 Monthly mean rainfall, Australia 1961–1990 (grey bars) and 2019 (green bars). Australia Bureau of Meteorology Annual Climate Statement (2019).

formal attribution study (van Oldenborgh et al. 2020) found that anthropogenic climate change increased the probability of extreme heat, such as that observed in 2019/2020, by more than two-fold. But it was not possible to identify an unambiguous climate change effect on drought, given the strong, short-term variations in 2019 in naturally occuring climate systems (e.g., the Indian Ocean Dipole).

Extreme weather events, such as droughts, floods and heatwaves, may have direct damaging effects on mental as well as physical health. Using a regular nationwide population survey that included questions about self-perceived mental health, Obradovich et al. (2018) estimated exposure to Hurricane Katrina increased the prevalence of mental health issues by about 4%. The same study found a link between rising temperatures between 2002 and 2012 and more frequent reporting of mental health difficulties. In the Pacific, sea-level rise and changes in storm frequency threaten long-term habitability of many low-lying islands, and challenge the basis of good mental health. The distress experienced by these populations manifests in many, culturally-nuanced ways (Tiatia-Seath et al. 2018).

It is projected that climate change will limit the amount and quality of freshwater resources in many places, due to changes in rainfall, higher temperatures, vegetation die-off and sea-level rise. Other factors such as population growth will influence the levels of freshwater stress, but climate variables dominate in many regional analyses (Karnauskas et al. 2018). When freshwater sources are compromised, the frequency of water-borne infectious diseases typically increases (Hodges et al. 2014). And there may be effects also on some chronic diseases. In coastal parts of Bangladesh, increasing frequency of storm surges associated with sea-level rise and changes in river flows have led to increased sea-water intrusions into aquifers used for drinking water. The increased intake of sodium in the population is linked strongly with high blood pressure, and increases the risk of cardiovascular diseases (Scheelbeek et al. 2017).

Since AR5, the growing evidence of indirect effects of climate change on health via ecosystem disturbance includes threats to food security and nutrition. Rising temperatures and shifting rainfall will limit or diminish crop yields in some parts of the world, and stunting (the principal consequence of child undernutrition) will follow in poor populations unable to manage rising food prices (Lloyd et al. 2018). Even in high-income countries, it may be difficult to maintain food production at healthy levels if high-impact climate 'tipping points' are crossed. In one analysis, examining the potential collapse of the Atlantic Meridional Overturning Circulation, the viability of arable farming in much of Great Britain would be in question (Ritchie et al. 2020).

Vector-borne diseases remain a potent health threat where changing climate intersects with fragile public health services. Recent outbreaks of dengue in Nepal and Banglagesh are a reminder. Studies that attempt to model the future risk of dengue worldwide, project a substantial increase in the number of people potentially affected. For example, Messina et al. (2019) estimate more than 2 billion extra people would be at risk in 2080. The biggest rises are likely to occur in parts of the world where dengue is already common, and future population growth and social development will strongly modify the effects of climate.

Health-damaging social disruption resulting from extreme weather is well-demonstrated by the consequences of severe storms (Woodward and Samet 2018). While it is still uncertain how the frequency of hurricanes will be affected by climate change, there is little doubt that a hotter, moister atmosphere will promote storms of greater intensity. Hurricane Maria caused widespread damage in Puerto Rico in 2017, and according to the official records, there were 64 deaths associated directly with the storm. But a survey in the three months after the hurricane found the mortality rate was increased by 62%, with approximately 4600 deaths occurring over and above what was expected (Kishore et al. 2018). Contributing factors included lack of shelter and suitable housing, interruptions to supplies of food and safe water, and breakdowns in social care and support for the elderly and chronically ill. One-third of the deaths were attributed to delayed or disrupted healthcare (Figure 3.2).

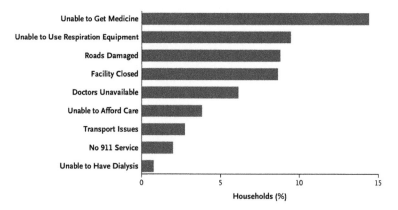

Figure 3.2 Proportion of households experiencing difficulties with access to medical services in the three months after Hurricane Maria, Puerto Rico 2017. (Kishore, N. et al. 2018).

3.4   Transition Risks

'Transition risk' refers to the collateral damage that might result from efforts to manage climate change. The changes that must be made to meet targets such as those agreed in Paris in 2015 will inevitably be disruptive. There are many paths to the 2 °C goal, but all of them require that global greenhouse emissions reach a peak very soon, and then reduce rapidly to reach zero or negative emissions around 2100.

Here are two examples of transition risks that are important to health and might be encountered if emissions enter free-fall within a decade or two. It seems unlikely that transformative change will be made without putting a price on carbon. Exactly what the price will be is not known of course. US\$200 a ton is towards the higher end of what is thought necessary, but not implausible. In 2015, emissions from the UK National Health Service amounted to almost 23 million tons of carbon dioxide equivalent (MtCO$_2$e). At \$200 a ton, the carbon liability amounts to about £3 billion, almost 20% of the total spend each year in the UK on medicines (Torjesen 2016). It is likely that a transfer of this magnitude, if it was imposed abruptly, would cripple any health system. But it must be noted also that the UK health service has been a model in many ways, achieving year-on-year reductions in greenhouse emissions despite increasing health service activity (Pencheon 2018).

The longer the emissions peak is postponed, and the shallower the decline in emissions subsequently, the greater the reliance on negative carbon technologies to meet climate targets. This means finding ways of drawing carbon out of the atmosphere and storing it securely. Planting trees is no more than a short-term solution, but it is the only carbon capture mechanism that is tested, well-understood and suitable for application at scale. The greater the reliance on forests for climate control, the more intense the competition for land that is presently used for other purposes. If emissions do not peak until after 2025, and the overshoot late century is managed substantially by land-based carbon capture, this may require more than half the available cropland area, worldwide (Obersteiner et al. 2018). How could this be managed without jeopardising food security, especially in countries where millions of people live with restricted incomes and there is little extra land available for agricultural expansion (Hasegawa et al. 2015)? Note that the relatively modest diversion of maize and sugar cane to biofuels in the 2000s was responsible for perhaps a third of the increase in global food prices that occurred during that decade (Rosegrant 2008).

3.5   Co-benefits – They are There, but Cannot be Assumed

Sometimes radical change is welcome because it provides opportunities that would not occur otherwise. It has been claimed that climate change is 'not just a challenge, but the greatest public health opportunity of the 21st century' (Watts et al. 2015). Is this the smooth road ahead? Are there chances here to avoid the health impacts of climate change without causing pain or loss? In my view, the 'look for opportunities' approach is important, but should not be over-sold.

Better housing illustrates how multiple benefits may be achieved by wise policy. In New Zealand the climate is mild, but many houses are drafty, cold and damp, as a result of poor design and unsatisfactory construction. Community trials of retrofitted insulation and clean, efficient heating have shown that it is possible to simultaneously raise temperatures indoors, improve health outcomes (fewer days off school for children with asthma, for instance) and lower power consumption (Howden-Chapman et al. 2008). Insulation alone provided total benefits in 'present value' (discounted) terms that were one and a half to two times the cost of the intervention (Chapman et al. 2018).

The bicycle, originally known as the 'pedestrian accelerator', shrinks cities and improves access to desired destinations, it democratises transport, and rewards physical effort (Zapata-Diomedi et al. 2017). The bicycle promotes frugal use of natural resources, including fossil fuels. A study of the co-benefits of investing in a cycling infrastructure in Auckland, New Zealand, found there was a cumulative net benefit of hundreds of millions of NZ dollars taking into account air pollution, car crash injuries, physical inactivity, fuel costs to households and greenhouse gas emissions (Macmillan et al. 2014).

There may be substantial gains from promotion of local, plant-based diets. According to Springmann et al. (2016), a carbon tax on food might each year save a gigaton of emissions (10^9 tons of carbon dioxide equivalents) and avoid up to 500,000 deaths globally. Simple dietary changes in India, replacing highly refined foods with alternatives such as pulses, coarse cereals and dark green vegetables, could reduce both dietary deficiencies that are now prevalent and cut greenhouse emissions from agriculture (Rao et al. 2018).

It makes sense to link warnings about future climate change with positive signals about what can be done immediately. It appears intuitively correct that 'alert' must be followed by 'rescue' if action is desired. But there are also reasons to be cautious. Climate priorities and needs may coincide with what is required to protect and promote health, but they may not. There is no necessary alignment. The move away from heavy reliance on coal as a primary fuel has simultaneously saved greenhouse emissions and protected human health. However, the promotion of diesel as an efficient motor fuel was good for the climate (due to less tailpipe CO_2), while increasing air pollution deaths and morbidity from toxic emissions such as nitrogen oxides (NO_x) and particulates.

It may seem obvious that knowledge of co-benefits will spur action to control climate change, but this is not necessarily so. Demonstrating multiple benefits is not the same as persuading. It has been evident for some time, sceptics might argue, that bicycles are good for the planet and they improve people's health, but there is no sign, in most cities, that fewer people are driving cars. Will more science, that quantifies the multiple benefits even more precisely, make a difference?

3.6   Conclusion

Climate change belongs in a new category of global environmental health problems. It is not just that the impacts are widely distributed; climate change is a result of unbalanced global systems. It is one of the modern threats to a 'safe operating space' for the planet. The effects on health occur directly, via floods and droughts, for example; through pressures on natural systems (reduced crop yields and undernutrition, for instance); and as a consequence of social disruption. Also, policy responses to climate change may affect health, intentionally or not; these are so-called 'transition risks'. Improving baseline health status is fundamental to coping with climate change, because the populations that are most seriously affected are those that already bear a heavy burden of disease. But an undifferentiated public health response is not sufficient. There are distinctive features of climate change that have to be taken into account. Mitigation, or primary prevention, will require rapid, deep cuts in greenhouse emissions if global heating is to be limited. The goal is to identify common solutions, responses to climate change that are health-enhancing rather than health-damaging. There are many candidates, but by and large they are not on the path of 'business as usual' development.

Acknowledgement

This chapter draws on the Redfern Oration, delivered to the Annual Congress of the Royal Australasian College of Physicians in Sydney in May 2018, and published subsequently (Woodward 2019).

References

Australia Bureau of Meteorology Annual Climate Statement (2019). http://www.bom.gov.au/climate/current/annual/aus/?utm_medium=social%20media&utm_source=twitter&utm_campaign=climate&utm_term=climate&utm_content=video-080120-annualclimatestatement#tabs=Rainfall

Bennett, C.M., Dear, K.B.G., and McMichael, A.J. (2013). Shifts in the seasonal distribution of deaths in Australia, 1968–2007. *International Journal of Biometeorology* April 24.

Borchers Arriagada, N., Palmer, A.J., Bowman, D.M. et al. (2020). Unprecedented smoke-related health burden associated with the 2019/20 bushfires in eastern Australia. *The Medical Journal of Australia* 2020 Mar 12.

Callaghan, M.W., Minx, J.C., and Forster, P.M. (2020). A topography of climate change research. *Nature Climate Change* 1, 10 (2): 118–123.

Chapman, R., Keall, M., Howden-Chapman, P. et al. (2018). A cost benefit analysis of an active travel intervention with health and carbon emission reduction benefits. *IJERPH* 15 (5): 962–10.

Di Virgilio, G., Evans, J.P., Blake, S.A.P. et al. (2019). Climate change increases the potential for extreme wildfires. *Geophysical Research Letters* 24, 46 (14): 8517–8526.

Hanigan, I.C., Dear, K., and Woodward, A. (2021). Increased ratio of summer to winter deaths due to climate warming in Australia, 1968–2018. *Aust NZ Journal of Public Health.* doi:10.1111/1753-6405.13107.

Hasegawa, T., Fujimori, S., Shin, Y. et al. (2015). Consequence of climate mitigation on the risk of hunger. ACS Publications. *American Chemical Society* Jun 2.

Hodges, M., Belle, J.H., Carlton, E.J. et al. (2014). Delays in reducing water-borne and water-related infectious diseases in China under climate change. *Nature Climate Change.* Nov 2.

Howden-Chapman, P., Pierse, N., Nicholls, S. et al. (2008). Effects of improved home heating on asthma in community dwelling children: Randomised controlled trial. *BMJ* 337: a1411.

Karnauskas, K.B., Schleussner, C.-F., Donnelly, J.P. et al. (2018). Freshwater stress on small island developing states: Population projections and aridity changes at 1.5 and 2 °C. *Regional Environmental Change.* 8, 18 (8): 2273–2282.

Kishore, N., Marqués, D., Mahmud, A. et al. (2018). Mortality in Puerto Rico after Hurricane Maria. *The New England Journal of Medicine* 29: NEJMsa1803972.

Lai, H., Hales, S., Woodward, A. et al. (2020). Effects of heavy rainfall on water-borne disease hospitalizations among young children in wet and dry areas of New Zealand. *Environment International* 145: 106136–5.

Lee, D., Chang, H.H., Sarnat, S.E. et al. (2019). Precipitation and Salmonellosis incidence in Georgia, USA: Interactions between extreme rainfall events and antecedent rainfall conditions. *Environmental Health Perspectives.* 4th Ed., 127 (9): 097005–12.

Lloyd, S.J., Bangalore, M., Chalabi, Z. et al. (2018). A global-level model of the potential impacts of climate change on child stunting via income and food price in 2030. *Environmental Health Perspectives* 126 (9): 097007–15.

Macmillan, A., Connor, J., Witten, K. et al. (2014). The societal costs and benefits of commuter bicycling: Simulating the effects of specific policies using system dynamics modeling. *Environmental Health Perspectives* 122 (4): 335–344.

Messina, J.P., Brady, O.J., Golding, N. et al. (2019). The current and future global distribution and population at risk of dengue. *Nature Microbiology* 3: 1–10.

Nolan, R.H., Boer, M.M., Collins, L. et al. (2020). Causes and consequences of eastern Australia's 2019//20 season of mega-fires. *Global Change Biology* 22, 58: 233–3.

Obersteiner, M., Bednar, J., Wagner, F. et al. (2018). How to spend a dwindling greenhouse gas budget. Nature climate change. *Nature Publishing Group* 1, 8 (1): 7–10.

Obradovich, N. (2018). Empirical evidence of mental health risks posed by climate change. *PNAS* 23, 115 (43): 10953–10958.

Pachauri, S., Van Ruijven, B.J., Nagai, Y. et al. (2013). Pathways to achieve universal household access to modern energy by 2030. *Environmental Research Letters* 2, 8 (2): 024015–8.

Pandey, B.D. and Costello, A. (2019). The dengue epidemic and climate change in Nepal. *The Lancet* 14, 394 (10215): 2150–2151.

Pencheon, D. (2018). Developing a sustainable health care system: The United Kingdom experience. *The Medical Journal of Australia* 208 (7): 284–285.

Rao, N.D., Min, J., and DeFries, R. (2018). Healthy, affordable and climate-friendly diets in India. *Global Environmental Change* 1, 49: 154–165.

Ritchie, P.D.L., Smith, G.S., Davis, K.J. et al. (2020). Shifts in national land use and food production in Great Britain after a climate tipping point. *Nature Food* 8: 1–18.

Rosegrant, M. (2008) Biofuels and grain prices: Impacts and policy responses. Testimony for the U.S. Senate Committee on Homeland Security and Governmental Affairs May 7, 2008 http:// large.stanford.edu/courses/2011/ph240/chan1/docs/rosegrant20080507.pdf

Scheelbeek, P.F.D., Chowdhury, M.A.H., Haines, A. et al. (2017). Drinking water salinity and raised blood pressure: Evidence from a cohort study in coastal Bangladesh. *Environmental Health Perspectives* 30, 125 (5): 1–8.

Smith, K.R., Frumkin, H., Balakrishnan, K. et al. (2013). Energy and human health. *BMJ* 34: 159–188.

Springmann, M., Mason-D'Croz, D., Robinson, S. et al. (2016). Global and regional health effects of future food production under climate change: A modelling study. *The Lancet* 1–10.

Tiatia-Seath, J., Underhill-Sem, Y., and Woodward, A. (2018). The Nexus between climate change, mental health and well-being and Pacific peoples. *PacHealthDialog* 30, 21 (2): 47–49.

Torjesen, I. (2016). NHS spent 8% more on medicines last year. *British Medical Journal* 22: i6320.

Van Oldenborgh, G.J., Krikken, F., Lewis, S. et al. (2021). Attribution of the Australian bushfire risk to anthropogenic climate change. *Natural Hazards and Earth System Sciences* 21 (3): 941–960.

Vardoulakis, S., Jalaludin, B.B., Morgan, G.G. et al. (2020). Bushfire smoke: Urgent need for a national health protection strategy. *The Medical Journal of Australia* 23. doi:10.5694/mja2.50511.

Watts, N., Adger, W.N., Agnolucci, P., et al. (2015). Health and climate change: Policy responses to protect public health. *Lancet* (published online June 23. .

Woodward, A. (2019). Climate change: Disruption, risk and opportunity. *Global Transitions* 1: 44–49.

Woodward, A., Smith, K.R., Campbell-Lendrum, D. et al. (2014). Climate change and health: On the latest IPCC report. *Lancet* 383 (9924): 1185–1189.

Woodward, A.J. and Samet, J.M. (2018). Climate change, hurricanes, and health. *American Journal of Public Health* 08 (1): 33–35.

Zapata-Diomedi, B., Knibbs, L.D., Ware, R.S. et al. (2017). A shift from motorised travel to active transport: What are the potential health gains for an Australian city? *PLoS ONE Public Library of Science*, 12 (10): e0184799.

4

Flooding and Public Health in a Changing Climate

Owen Landeg

National Institute for Health Protection, London School of Hygiene and Tropical Medicine

4.1 Introduction

Floods are the most frequent disaster worldwide. The public health consequences of flooding are significant and long-lasting. Often, only the immediate, physical risk to life is considered during the emergency response period of a flood, resulting in the broader public health and well-being impacts, or those caused by infection or chemical hazards, being poorly understood. It is not always easy to identify the longer-term health consequences of flooding, such as effects caused by secondary stressors (i.e., factors indirectly related to the event itself that can cause mental health impacts, such as displacement, destruction of homes, delayed recovery).

The actions taken before, during and after a flood event considerably influence the short- and long-term health outcomes. Public Health is therefore an important advocate across the multi-disciplinary policy and operational landscapes for increasing flood resilience in a way that avoids unintended health consequences and exacerbating inequality.

4.2 Types of Floods

Flooding can be defined as an overflow of a large amount of water beyond its normal limit, especially over what is normally dry land. Fluvial or river floods occur when a river, lake or stream rises and overflows its surrounding banks, particularly in response to significant or excessive rainfall. Tidal flooding can be defined as the temporary inundation of low-lying coastal areas, during exceptionally high tides, or a combination of high tides, a low pressure system and a coastal surge driving sea water on to land. Pluvial flooding, otherwise known as surface water or flash flooding, occurs when an extreme rainfall event creates a flood independent of an overflowing water body. Pluvial flooding can occur in rural or urban environments. Less common flood sources include the

Hydrometeorological Extreme Events and Public Health, First Edition. Edited by Franziska Matthies-Wiesler and Philippe Quevauviller.
© 2022 John Wiley & Sons Ltd. Published 2022 by John Wiley & Sons Ltd.

failure of infrastructure, for example dam collapse or flood barrier failure and anthropogenic sources such as a burst water mains pipe.

From a public health perspective, fast onset flood events can have more dangerous direct impacts on health (i.e., death, injury) than slow onset events, due to the rapid increase in water levels and the typical short time frame available for warning, informing and action.

4.3 Health Impacts of Flooding

The health impacts of flooding can be segregated according to their typical onset from exposure to the flood event (Menne and Murray 2013). However, it is important that public health also considers any concurrent and cascading risk associated with a flood (i.e., power outages, water supply disruption) and the need to amend the public health response strategy accordingly. Furthermore, concurrent or cascading risks are likely to influence the health impacts of a flood event and therefore require additional or adapted interventions.

Box 4.1 *Immediate and early health effects of flooding*

- drowning
- physical trauma from concealed or displaced objects
- health effects of water shortages and contamination due to loss of water treatment works
- health effects of chemical contamination of flood water
- heart attacks, electrocution, fire
- distress

Medium to longer-term health effects of flooding

- carbon monoxide poisoning when petrol or diesel generators or other similar fuel-driven equipment are used indoors for drying or pumping out flood water, cooking, lighting or heating
- infectious disease outbreaks (water-, rodent- or vector-borne)
- illness associated with disruption and reduced access to healthcare services
- anxiety, depression and post-traumatic stress disorder (PTSD)

4.3.1 Populations at Risk

While a whole population is at risk of the health consequences of flooding, certain groups may be more susceptible. Vulnerability to the health effects of flooding is transient, complex and determined by factors beyond the magnitude of the event. These dynamic factors include health and social circumstances and behavioural choices. Some groups, who may not be thought of as vulnerable (i.e., young men), may put themselves at risk by driving through flood water for example. Moreover, some people may be in a temporary state of risk at the time of flooding, for example, those recently discharged from hospital.

The following individuals and communities are considered at greater risk of experiencing the negative health and well-being impacts from flooding:

- older people
- people with physical, sensory and cognitive impairments
- people with chronic diseases
- those receiving care at home (i.e., home oxygen, dialysis, palliative care)
- children
- pregnant women
- people with language and cultural-based vulnerabilities
- people who are homeless
- transient communities (i.e., tourists, migrant workers)
- people who are socially isolated or have weak social networks.

4.3.2 Mortality

It is estimated that two-thirds of deaths associated with flooding during the emergency response phase are from drowning. Other causes of death include physical trauma, heart attacks, electrocution, hypothermia and fire. The highest death tolls are typically associated with flash floods. Fatalities in Higher Income Countries (HICs) are most associated with people in motor vehicles. The number of flood-related fatalities are significantly higher in Lower to Middle Income Countries (LMICs) compared to HICs, because LMICs typically have populations more susceptible to the health consequences of flooding, have larger populations and have less resilient infrastructure (i.e., water, sanitation, power). The Centre for Research on the Epidemiology of Disasters estimates the ratio of flood-related mortality in LMICs vs. HIC worldwide is almost 23 to 1.

Pathogens do not survive in human bodies after death, therefore the epidemiological risk from dead bodies is limited, especially when public health is followed. In HICs it is unlikely that the number of fatalities that would occur during the acute phase would exceed normal local arrangements. However, if needed, organisations would need to draw upon mass fatality plans. Any fatality plans should be integrated within the wider multiagency flooding response plan (see Chapter 6).

Regardless of the context, it is important that the basic values such as respect for the dead and the right of survivors to properly conduct a funeral are observed where possible. Special attention should be given to providing personnel with protection equipment against blood-borne viruses and other infections for those handling corpses. Where a flood occurs during an outbreak of certain infectious diseases, with the risk for potential transmission from dead bodies, special measures may be required.

4.3.2.1 Exacerbation of Pre-Existing Conditions
Flooding can significantly disrupt the healthcare system and affect healthcare delivery. The exacerbation of pre-existing health conditions represents a burden on public health that can lead to complications or even death. Flooding can cause reduced access to healthcare as transport supplies are affected in the short term, disruption to prescription medication can occur through

evacuation or disruption to medical supply chains and routine care can be hindered. A study of British diabetics in the year following the 2007 English floods found that their glycaemic control had deteriorated amongst users of insulin (Ng et al. 2011).

Those receiving care for long-term conditions may be disrupted by flooding. This is particularly concerning in HICs, where changes to the models of care mean more and more complex care is being delivered in the residential setting rather than in hospital. For example, patients requiring dialysis are at great risk of treatment disruption and exacerbation, as dialysis requires the provision of power and clean water. It is important that patients register for Priority Service Registers where available, so utility companies are aware of their vulnerability and can appropriately prioritise their reconnection whilst supporting the individual.

Flooding can also cause a deterioration of living conditions, with mould and damp commonly being a long-term impact in residential houses. Coupled with disruption to water supplies, extreme temperatures and fuel poverty, patients with chronic lung conditions for example, can see their symptoms worsened by cold, stress and fatigue.

4.3.3 Flooding and Mental Health

Flood-related mental health problems are a significant health burden irrespective of the country context. However, psychological morbidity is typically one of the major public health impacts within England (Jermacane et al. 2018). Although several studies have described the mental health impacts of flooding using qualitative methods or cross-sectional surveys in the first months after a flooding event, there is very limited evidence available to help us understand the scale, intensity or duration of psychological morbidity. Better information is needed to support those affected and to inform decisions about services and interventions before, during and after a flood.

The storms of the winter 2013/14 brought the wettest winter to England in 20 years (Kendon and McCarthy 2015). From December 2013 through to February 2014, a series of 12 major storms caused record levels of rainfall, river flows, sea levels and wave heights and communities faced widespread flooding. Approximately 11,000 properties were flooded from December 2013 to May 2014 and 155 severe flood warnings (severe flooding, danger to life) were issued.

Public Health England's (PHE) response to the floods included the establishment of the English National Study of Flooding and Health working with academic partners to investigate the long-term health impacts of flooding and related ongoing disruptions on mental health (Waite et al. 2017).

The National Flooding and Health Study in England found that the mental health impacts of flooding are prolonged and include not just those who are directly flooded (i.e., flood water within a home), but those who live in the vicinity of flooded homes or in a community affected by the flood event (Waite et al. 2017). High levels of probable depression, anxiety and PTSD were reported 12 months after the winter 2013/14 England floods by participants whose homes had been flooded (Table 4.1). Elevated levels were also found in those participants disrupted by the flooding, but not directly flooded. Disruption to health and social care services was found to increase the risk of poor mental health outcomes for those who routinely use these services.

Secondary stressors can be defined as those sources of stress not directly associated with flood water, but nonetheless influence the psychological morbidity following a flood event. The typology of secondary stressors includes displacement, loss of utilities (e.g., water, sewage,

Table 4.1 Prevalence of mental health outcomes by exposure group (after Waite et al. 2017).

Outcome	Overall cohort	Unaffected	Exposure group	
			Disrupted	Flooded
Probable depression	250/1929 (12.6%)	16/278 (5.8%)	102/1058 (9.6%)	125/593 (20.1%)
Probable anxiety	300/1927 (15.6%)	18/278 (6.5%)	113/1052 (10.7%)	169/597 (28.3%)
Probable PTSD	396/1925 (20.6%)	22/278 (7.9%)	160/1056 (15.2%)	214/591 (36.2%)

power, communication), disrupted healthcare (access difficulties, loss of medicines or devices), disrupted work or education, loss of possessions (i.e., sentimental items) and difficulties with recovery (repair, insurance), loss of amenity or even livelihood (Lock et al. 2012). These should be factored into the long-term response within the reconstruction phase following a flood. The incorporation of a single point of contact for insurance claims embedded within the local flood response has served to aid claimants navigating the claims process.

Flooding is likely to exacerbate the challenge of poor mental health in many communities. Commissioners and providers of mental health services as well as emergency planners should prepare for an increased need for services in areas affected, or likely to be affected, by flooding. The World Health Organization's (WHO) recommended approach of psychological first aid in the initial aftermath of flooding should be taken to avoid pathologisation of the normal distress caused by a flood (WHO Regional Office for Europe 2014). At a population level, referral for specialist help should only happen if distress does not resolve after several weeks.

4.3.4 Flooding and Infectious Diseases

Flood water is likely to be contaminated by disease producing bacteria and viruses, but not high-risk enteric infectious diseases (e.g., cholera, typhoid), which are not naturally endemic in the UK and other HIC populations. The risk to people in HICs from bacterial contamination of flood water is, therefore, relatively low, if public health advice is followed (e.g., hand washing and wearing rubber boots and gloves) and assuming resilient water treatment and sanitation infrastructure. Where there is any raw sewage entering flood water, the diluting and dispersing of potential sources of infection further significantly reduces any risk. For these reasons, infectious disease outbreaks after a flood are much more likely in LMICs than HICs.

4.3.4.1 Water-Borne Diseases   Water-borne diseases are those caused by pathogenic microorganisms spread by faecal-oral transmission when drinking water is contaminated by human or animal excreta. Pathogens include bacteria, viruses, protozoa and helminths. While some diseases such as hepatitis E, typhoid fever or cholera are confined to certain tropical countries, others such as *Campylobacter, Shigella, Cryptosporidium* and *Rotavirus* are more widespread. Most pathogens cause diarrheal diseases; however, some cause non-diarrheal disease such as typhoid.

Exposure and the subsequent outbreak of water-borne diseases can be exacerbated by disruption to water treatment and sanitation. Those populations with weak water and sanitation infrastructure or poor hygiene before the onset of flooding are particularly at risk. Overcrowding, poor housing conditions and low socioeconomic status also serve to increase the likelihood of transmission.

4.3.4.2 Respiratory Infections Increased incidence of respiratory illness (i.e., upper respiratory tract infections) has been recorded following flooding. Social and housing aspects serving to increase the transmission of pathogens and individual vulnerability to respiratory infections include overcrowding, poor ventilation and damp or mould conditions. Environmental factors such as cold weather can lead to fuel poverty. Susceptibility to disease is further influenced by impaired immunity linked to undernutrition, stress and other factors.

4.3.4.3 Vector-Borne Diseases Flooding or heavy rainfall can lead to an increase in the breeding sites for mosquito-borne diseases in those areas where such diseases are endemic. This has been known to significantly increase malaria incidence and seasonal Rift Valley Fever in Eastern Africa. Large outbreaks of dengue haemorrhagic fever and chikungunya have also been reported following flooding.

In the initial days after a flood, existing larvae and breeding sites can be washed away, which can result in an initial decrease in vector-borne diseases, with incidence increasing as flood water recedes, thus re-establishing breeding sites. Rising temperatures have also been found to influence prevalence. Other studies, however, found no change in larvae and adult vector populations in response to temperature and rainfall changes, meaning factors related to population dynamics may be more influential on inter-epidemic disease variation than meteorological conditions.

It is important to recognise that mosquito-borne diseases do not all share the same characteristics, including ecology and patterns of transmission, and external factors such as temperatures, seasonal variation and disease dynamics need to be considered. For these reasons, the risk posed by vector-borne diseases following flooding needs to be evaluated on an individual basis.

4.3.4.4 Rodent-Borne Diseases Increases in rodent-borne diseases have been reported following flooding. This is due to increasing rodent populations, increased contact between humans and rodents, particularly in the cleaning-up recovery phase of a flood, and the transmission of pathogens via flood water. Leptospirosis, caused by the shedding of the *Leptospira interrogans* pathogen by rodents in their urine, is the main rodent-borne disease threat posed by flooding. Education and targeted awareness campaigns for specific risk groups is therefore required.

4.3.4.5 Skin Infections Skin infections can result following flooding, through contact with flood water or contaminated soil or surfaces. Compromised hand hygiene or failure to treat wounds can lead to severe infections, even if the initial wound was minor. The risk of skin infections is minimised where public health advice including the importance of wearing rubber gloves, boots and other protective equipment is followed.

4.3.5   Displacement, Evacuation and Sheltering

Flooding, particularly severe flooding, can lead to significant population displacement which can hinder public health surveillance, disrupt existing public health programmes (i.e., vaccinations) and lead to overcrowding, reduced access to healthcare, and compromised access to safe drinking water and sanitation. The movement of people after a flood can also remove people from their community networks, sometimes for significant periods of time.

During the acute phase of a flood, the urgent evacuation of people without warning has been linked to cardiovascular events such as heart attacks. The establishment of shelters can cause a disruption of healthcare provision, with people evacuating without their prescription medication. The identification and sourcing of medication can be challenging within an evacuation shelter setting (Ochi et al. 2014), but the presence of prescribing nurses has been known to mitigate the health risk of disrupted healthcare delivery. Evacuation also presents challenges for the tracking of patients receiving routine treatment and the identification of vulnerable at-risk adults and children.

Lack of adherence to evacuation orders, or failure of provisions at evacuation centres, can lead to people refusing to evacuate. For example, cases of people refusing to attend evacuation centres due to lack of provision for pets have been reported. Cases of people refusing to evacuate within the community have also been reported. In these situations, volunteers, particularly spontaneous volunteers, have been known to not accept, or question, an individual's right to refuse to evacuate. In these instances, training on mental capacity and the rights of the individual within the country context is required.

4.4   Health System Resilience

According to the UK Health Accounts, published by the Office for National Statistics, healthcare expenditure in the UK accounted for 10% of gross domestic product in 2018. Ascertaining the current level of risk is challenging due to the dynamic nature of flood risk and complexity of the health system, including health assets (i.e., hospitals) and the diffuse nature of some healthcare delivery such as domiciliary care.

Many ambulance stations, fire stations, police stations, hospitals, care homes and surgeries in the UK are located in areas that are susceptible to fluvial and coastal flooding.

According to the Environmental Agency, approximately 13% of ambulance stations, 7% of hospitals and 9% of surgeries in England are located in areas that are susceptible to fluvial and coastal flooding (Environmental Agency 2009). Assuming a continuation of climate change, future climate projections for the UK suggest that flood risk to healthcare assets is likely to increase. The proportion of care homes, emergency services, hospitals and GP surgeries at risk of flooding in England is set to increase by 13%, 11%, 4% and 12%, respectively, by the 2050s under a 2-degree warming scenario.

Future projections indicate an increase in the number of GP surgeries, care homes, emergency service stations and hospitals in the flood risk zones, with the largest change in risk generally shown for care homes (medium magnitude, low confidence). By the 2050s, under a 4-degree scenario, the number of hospitals in England located in areas at medium or high risk of flooding (1 in 200 annual chance of flooding or greater) increases to between 187 and 200 and the number of care homes increases to between 1,338 and 1,454. The projections above assume no population growth, and the ranges are across the different adaptation scenarios considered in the CCRA (CCRA 2017).

The resilience of healthcare systems is an emerging topic of international importance, reflecting concerns about the wide-ranging consequences for human health from climate change. Healthcare systems themselves need to be resilient to climatic events to ensure that there is sufficient capacity to address the impacts of climate change on human health whilst maintaining healthcare delivery. The WHO's working definition of a climate resilient health system is one 'that is capable to anticipate, respond to, cope with, recover from and adapt to climate related shocks and stress, so as to bring sustained improvements in population health, despite an unstable climate.' WHO (2015) Operational framework for building climate resilient health systems.

Storms and floods are known to reduce the ability of healthcare systems to respond to health crises, affecting the quality of healthcare provision. Despite recognition of the vulnerability, there remains a lack of research into how climate change will impact on healthcare systems and what mitigation measures are available. Furthermore, the resilience of health infrastructure in a future climate remains largely undefined. In the context of the present time, few studies have explored in detail the impacts of flooding on healthcare systems, despite previous flood events having detrimental effects on healthcare system infrastructure across Europe.

Work assessing the healthcare system impacts associated with the December 2013 east coast storm surge and subsequent flooding in Boston, a port town in Lincolnshire, England, demonstrated the difficulties experienced by healthcare services in responding to the flood event and gaps in preparedness (Landeg et al. 2019).

Flooding and coastal change risk are considered as one of the top six interrelated climate change risks for the UK, and analysis suggests that 0.5 to 1 m in sea-level rise could make some 200 km of coastal flood defences in England highly vulnerable to failure in storm conditions (ASC 2016). This indicates the scale of strategic prevention required to protect health in light of a changing climate. Healthcare leaders should consider future climates within the planning and development stage of new healthcare facilities, including flooding and other environmental hazards such as heatwaves. Assurance that health service providers are registered to directly receive high-impact weather alerts and know what actions to take upon receipt of the alert is required. With increasing fragmentation of health services in England and across the globe, greater vertical and horizontal communication is required to prevent siloed working across stakeholders including non-health partners.

Improved recording of flood impacts and disruption is required to demonstrate the ongoing economic cost of high-impact weather upon the global health sector. The wider emergency planning community should ensure that multiagency exercises contain a realistic portrayal of capacity within the health sector and consider concurrent risks such as the annual winter pressures. The creation of new exercises that accurately reflect the healthcare disruption should be considered. Measures and efforts to increase preparedness in the healthcare system need to be proportionate to the exposure of the system to weather-related hazards. More work is required to increase the climate resilience of the healthcare sector, particularly when climate change is set to increase the risk of weather-related impacts.

4.5   Flooding and Climate Change

Flooding and coastal change risks to communities, businesses and infrastructure is one of the top six interrelated climate change risks for the UK, with the Committee on Climate Change concluding that more action is required (ASC 2016). Flooding is also the highest environmental risk featured in the UK National Risk Assessment (UK Government National Risk Register 2020), in terms of impact and likelihood. This is replicated across the globe.

Human-induced global warming has already caused multiple observed changes in the climate system (Hoegh-Guldberg et al. 2018). Increasing sea levels and an increase in the frequency and magnitude of flooding are some of the most likely consequences of climate change. Despite global efforts in climate change mitigation and ever-increasing ambitious Net Zero targets, an increase in future flood risk will still occur due to the changes already made. For this reason, climate change adaptation strategies are a key component of our global climate change response in order to protect health and well-being. The impacts of climate change are likely to disproportionately affect the poorest communities, both nationally and internationally, and public health officials should consider the capacity or 'adaptation gap' of such individuals and communities to adapt and the support they may require. Furthermore, addressing the inequalities of climate change impacts and the opportunities presented by tackling climate change should be a key driver for public health action.

4.6   Public Health Mitigation, Planning and Prevention

Implementation of a multi-agency, all-hazards approach to emergency preparedness and strategic long-term prevention, translated into local action plans that include public health and the wider health system, is one of the most important measures to minimise the health impacts of flooding (see Chapter 6). This approach should have public health as a key partner in the long-term prevention action, the short acute response phase of a flood and the longer-term recovery efforts. Managing and preventing the health effects of flooding can be considered in three stages: primary, secondary and tertiary prevention.

Primary Prevention: These measures form part of the long-term planning for flooding and can be structural (e.g., engineering) or non-structural (policy and organisational). Examples include emergency plans, mutual aid arrangements and scenario exercising, promotion of flood alert service(s), land use management, tree planting, control of water sources and flow, property level flood protection, flood defences and barriers, design and architectural strategies and the availability of affordable flood insurance. During this phase, public health officials should be aware of potential long-term unintended health consequences, whilst simultaneously identifying opportunities for co-benefits. For example, green and blue spaces can provide flood protection whilst simultaneously improving air quality, reducing heatwave exposure and increasing active transport.

Secondary Prevention: These measures can be taken either just before or during a flood event to mitigate the health effects. Examples include the identification of at-risk or high-risk populations before floods occur (accounting for difficulties in communication and mobility and the needs of people with chronic diseases), early warning systems, evacuation plans including communication and information strategies, and planned refuge areas. It is important that consistent messages are promoted across organisations, with non-health partners incorporating and promoting key public health messages.

Tertiary Prevention: These measures can be taken during and after a flood to minimise health impacts. For example, moving belongings to safe areas, ensuring the provision of clean drinking water, health surveillance and monitoring of health impacts, treating ill people, and recovery and rehabilitation of flooded houses. Emergency responders should be aware of the 'recovery gap', the period after which an emergency response has ended, and people must rely on other sources of support for continued recovery. During this time several health, social and economic stressors may arise. Social care services should recognise that restoring communications and keeping families together are key measures to reducing suffering and promoting recovery after a flood. Healthcare providers should be aware of the long-term distress that flooding may cause for people who are affected.

4.7 Conclusions

Public Health is defined as the art and science of preventing disease, prolonging life and promoting health through the organised efforts of society. The health consequences of flooding are numerous and depend largely upon the magnitude and typology of the flooding event, the vulnerability of the affected population, and the country context. Public health is well placed to provide tailored specialist and technical advice to help ensure evidenced-based strategic prevention and an effective response and recovery which aid health protection (see Chapter 6).

With increasing exposure to flooding as a result of our changing climate, the burden on public health caused by flooding is set to increase. It is important that we continue to evaluate flood responses and develop the evidence base to strengthen national and international preparedness strategies that prevent ill health and promote resilient individuals and health services and communities.

References

ASC (2016). UK Climate Change Risk Assessment 2017 Evidence Report – Summary for England. Adaptation Sub-Committee of the Committee on Climate Change, London. Available at: https://www.theccc.org.uk/wp-content/uploads/2016/07/UK-CCRA-2017-England-National-Summary-1.pdf (accessed 21 October 2021).

CCRA (2017). https://www.theccc.org.uk/wp-content/uploads/2016/07/UK-CCRA-2017-England-National-Summary-1.pdf

Environment Agency (2009). Flooding in England: A National Assessment of Flood Risk. Available at: https://assets.publishing.service.gov.uk/government/uploads/system/uploads/attachment_data/file/292928/geho0609bqds-e-e.pdf (accessed 21 October 2021).

Hoegh-Guldberg, O.D., Jacob, M., Taylor, M. et al. (2018). Impacts of 1.5°C global warming on natural and human systems. In: *Global Warming of 1.5°C. An IPCC Special Report on the*

Impacts of Global Warming of 1.5°C Above Pre-industrial Levels and Related Global Greenhouse Gas Emission Pathways, in the Context of Strengthening the Global Response to the Threat of Climate Change, Sustainable Development, and Efforts to Eradicate Poverty (ed. V. Masson-Delmotte, P. Zhai, H.-O. Pörtner, et al.).

Jermacane, D., Waite, T.D., Beck, C.R. et al. (2018). The English National Cohort Study of Flooding and Health: The change in the prevalence of psychological morbidity at year two. *BMC Public Health* 18: 330. https://doi.org/10.1186/s12889-018-5236-9 Available at: https://bmcpublichealth.biomedcentral.com/articles/10.1186/s12889-018-5236-9#citeas (accessed 21 October 2021).

Kendon, M. and McCarthy, M. (2015). The UK's wet and stormy winter of 2013/2014. *Weather* 70 (2): Editorial. Available at: https://rmets.onlinelibrary.wiley.com/doi/pdf/10.1002/wea.2465 (accessed 21 October 2021).

Landeg, O., Whitman, G., Walker-Springett, K. et al.(2019). Coastal flooding and frontline healthcare services: Challenges for flood risk resilience in the English healthcare system. *Journal of Health Services Research & Policy* 24 (4): 219–228. doi: 10.1177/1355819619840672. Epub 2019 Jul 23. PMID: 31333054.

Lock, S., Rubin, G.J., Murray, V. et al. (2012). Secondary stressors and extreme events and disasters: A systematic review of primary research from 2010–2011. In: *PLOS Currents Disasters*. Edition 1. doi: 10.1371/currents.dis.a9b76fed1b2dd5c5bfcfc13c87a2f24f. Available at: http://currents.plos.org/disasters/index.html%3Fp=4597.html (acessed 21 October 2021).

Menne, B. and Murray, V. (2013). *Floods in the WHO European region: Health effects and their prevention*. Denmark: WHO Europe. http://www.euro.who.int/__data/assets/pdf_file/0020/189020/e96853.pdf. Available at: https://apps.who.int/iris/handle/10665/108625 (accessed 21 October 2021).

Ng, J., Atkin S.L., Rigby A.S. et al. (2011). The effect of extensive flooding in Hull on the glycaemic control of patients with diabetes. *Diabet. Med.* 28 (5): 519–524. doi: 10.1111/j.1464-5491.2011.03228.x. PMID: 21214625. Available at: https://pubmed.ncbi.nlm.nih.gov/21214625 (accessed 21 October 2021).

Ochi, S., Hodgson, S., Landeg, O. et al. (3014). Disaster-driven evacuation and medication loss: A systematic literature review. *PLoS Curr.* 2014:6:ecurrents.dis.fa417630b566a0c7dfdbf945910 edd96. doi: 10.1371/currents.dis.fa417630b566a0c7dfdbf945910edd96. Available at: https://www.ncbi.nlm.nih.gov/pmc/articles/PMC4169391 (accessed 21 October 2021).

UK Government. National Risk Register (2020). Available at: https://assets.publishing.service.gov.uk/government/uploads/system/uploads/attachment_data/file/952959/6.6920_CO_CCS_s_National_Risk_Register_2020_11-1-21-FINAL.pdf (accessed 21 October 2021).

Waite, T.D., Chaintarli, K., Beck, C.R. et al. (2017). The English national cohort study of flooding and health: Cross-sectional analysis of mental health outcomes at year one. *BMC Public Health* 17 (129). https://doi.org/10.1186/s12889-016-4000-2.

WHO Regional Office for Europe (2014). Floods and Health. Fact sheets for health professionals. Available at: https://www.euro.who.int/__data/assets/pdf_file/0016/252601/Floods-and-health-Fact-sheets-for-health-professionals.pdf (accessed 21 October 2021).

5

The Climate Change, Disaster Risk Reduction and Health Nexus

Demetrio Innocenti

University of Antwerp, Belgium

5.1 Introduction

This chapter focuses on a comparative analysis of the three major international frameworks that address development, Disaster Risk Reduction (DRR), and climate change in relation to their links with human health and well-being.

Specifically, it analyses the actions proposed under the Sendai Framework for Disaster Risk Reduction (SFDRR) (United Nations 2015a) and the Sustainable Development Goals (SDGs) (United Nations 2015b) and it will reflect on the current status of the incorporation of health in the Paris Agreement's Nationally Determined Contributions (NDCs) (UNFCCC 2015).

In its findings and conclusions, it discusses the current success stories and shortcomings in creating an effective health nexus amongst these three frameworks and proposes practical solutions on how to further integrate health into climate-related investments.

5.2 The Sendai Framework: Tackling Disaster Risk and Health at International Level

The SFDRR is a voluntary agreement adopted in March 2015 by 187 governments in Sendai, Japan.

In a previous volume of this series, the need for better integration between science and policy was emphasised (Innocenti 2015) to effectively address DRR policies and actions. That chapter was written back in 2014, in the context of a stronger

Hydrometeorological Extreme Events and Public Health, First Edition. Edited by
Franziska Matthies-Wiesler and Philippe Quevauviller.
© 2022 John Wiley & Sons Ltd. Published 2022 by John Wiley & Sons Ltd.

collaboration between universities, regional organisations such as the European Union (EU), and international players such as the United Nations Office for Disaster Risk Reduction (UNDRR). This collaboration is intended to provide policy-makers at national and international level with science-based evidence for improving the effectiveness of their policy (and consequently public spending in DRR).

The SFDRR (UN 2015a) is a good example of how the efforts to link science and policy in DRR and health were then translated into specific policy guidance. The 2015–2030 SFDRR is the successor of the 2005–2015 Hyogo Framework for Action (HFA) for DRR (UNISDR 2005) and, compared to its predecessor, it emphasises and better addresses the health dimension in disaster risk management policies, regulations and investments. This happened thanks to the consultations and inputs that the scientific community provided to the UNDRR (previously named UNISDR) in collaboration with partners such as the World Health Organisation (WHO) and the European Commission (EC). This led to a comprehensive international DRR framework, which shifts the perspective from managing disasters to managing risks.

Doing so, the SFDRR takes a truly holistic approach in advising governments and international donors on how to tackle cross-cutting issues such as health in the context of disaster prevention, preparedness and post-emergency.

The SFDRR counts more than 30 references to health (Kelman 2015). In addition, compared to the HFA, it mainly targets the nexus of epidemics and pandemics.

Investigating in further detail, the SFDRR presents seven targets and five priorities for action, which serve the following overreaching goal: 'The substantial reduction of disaster risk and losses in lives, livelihoods, and health and in the economic, physical, social, cultural and environmental assets of persons, businesses, communities, and countries.'

As noted by Aitsi-Selmi et al. and reported in Table 5.1, five of the seven SFDRR targets are relevant to health and well-being, and the language in the priorities of action reflects this (Table 5.2).

Table 5.1   The Sendai Framework seven targets.

SFDRR Seven Targets	Relevance to the health sector
A. Substantially reduce global disaster mortality	✓
B. Substantially reduce the number of affected people	✓
C. Reduce direct disaster economic loss in relation to global gross domestic product (GDP)	
D. Substantially reduce disaster damage to critical infrastructure and disruption of basic services	✓
E. Substantially increase the number of countries with national and local disaster risk reduction strategies by 2020	✓
F. Substantially enhance international cooperation to developing countries	
G. Substantially increase the availability of and access to multihazard early warning systems and disaster risk information	✓

Most important is the role of health in the four priorities of action of the SFDRR, namely:

- Priority 1. Understanding disaster risk
- Priority 2. Strengthening disaster risk governance to manage disaster risk
- Priority 3. Investing in disaster risk reduction for resilience
- Priority 4. Enhancing disaster preparedness for effective response and to 'Build Back Better' in recovery, rehabilitation and reconstruction

The relevance for the health sector was well mapped in the following paragraphs of the SFDRR, as reported in Table 5.2.

Table 5.2 Health and the Sendai Framework for Disaster Risk Reduction (adapted from Aitsi-Selmi A. et al., 2015).

SFDRR Priority and paragraph	Reference to health and well-being
Priority 3 (30i)	Enhance the resilience of national health systems: including by integrating disaster risk management into primary, secondary and tertiary healthcare, especially at the local level; developing the capacity of health workers in understanding disaster risk and applying and implementing DRR approaches in health work; promoting and enhancing the training capacities in the field of disaster medicine; and supporting and training community health groups in DRR approaches in health programmes, in collaboration with other sectors, as well as in the implementation of the 2005 International Health Regulations of the World Health Organization.
Priority 3 (30j)	Strengthen the design and implementation of inclusive policies and social safety-net mechanisms, including through community involvement, integrated with livelihood enhancement programmes, and access to basic healthcare services, including maternal, newborn and child health, sexual and reproductive health, food security and nutrition, housing and education, towards the eradication of poverty. To find durable solutions in the post-disaster phase and to empower and assist people disproportionately affected by disasters.
Priority 3 (30k)	People with life-threatening and chronic disease, due to their particular needs, should be included in the design of policies and plans to manage their risks before, during and after disasters, including having access to lifesaving services.
Priority 3 (31e)	Enhance cooperation between health authorities and other relevant stakeholders to strengthen country capacity for disaster risk management for health, the implementation of the International Health Regulations (2005) and the building of resilient health system.
Priority 4 (33c)	Promote the resilience of new and existing critical infrastructure, including water, transportation and telecommunications infrastructure, educational facilities, hospitals and other health facilities, to ensure that they remain safe, effective and operational during and after disasters in order to provide live-saving and essential services.
Priority 4 (33n)	Establish a mechanism of case registry and a database of mortality caused by disaster in order to improve the prevention of morbidity and mortality.
Priority 4 (33o)	Enhance recovery schemes to provide psychosocial support and mental health services for all people in need.

It is remarkable that in 2015, three major international frameworks due to shape the agenda of this decade have been all adopted by the large majority of the United Nations member countries: the SFDRR, Sustainable Development Goals (SDGs), and the Paris Agreement as the new pillar agreement to cope against the impact of a changing climate. A discussion on the convergencies (and divergencies) of these three major agreements would be outside the scope of this chapter. Yet, while the next section will elaborate further on the Paris Agreement and the linkage between climate change and health, it is possible to map the points of contacts between the SFDRR and the SDGs when it comes to ensuring that countries align their efforts on how to reduce disasters' impacts on people's health and well-being.

The SFDRR seems well integrated into the SDGs. The two frameworks align their reporting requirements, making the feedback and data analysis of the indicators of the countries to the UN and UNDRR more efficient, avoiding duplication of reporting efforts at the national level. In addition, both frameworks have the same lifespan of implementation (2015–2030).

Tables 5.3a, 5.3b and 5.3c give details on the alignment of the indicators between the SDGs and SFDRR. It is also highlighted where there is a connection to the health sector.

Table 5.3a   The Sustainable Development Goal 1 and the Sendai Framework interlinks.

SDG indicators	Goal 1. End poverty in all its forms everywhere	SFDRR indicators
1.5.2	Direct economic loss attributed to disasters in relation to global gross domestic product (GDP)	C1
1.5.3	Number of countries that adopt and implement national disaster risk reduction strategies in line with the Sendai Framework for Disaster Risk Reduction 2015–2030	E1
1.5.4	Proportion of local governments that adopt and implement local disaster risk reduction strategies in line with national disaster risk reduction strategies	E2

Table 5.3b   The Sustainable Development Goal 11 and the Sendai Framework interlinks.

SDG indicators	Goal 11. Make cities and human settlements inclusive, safe, resilient and sustainable	SFDRR indicators
11.5.1	Number of deaths, missing persons and directly affected persons attributed to disasters per 100,000 population	A1 and B1
11.5.2	Direct economic loss in relation to global GDP, damage to critical infrastructure and number of disruptions to basic services, attributed to disasters	C1, D1, D5
11.b.1	Number of countries that adopt and implement national disaster risk reduction strategies in line with the Sendai Framework for Disaster Risk Reduction 2015–2030	E1
11.b.2	Proportion of local governments that adopt and implement local disaster risk reduction strategies in line with national disaster risk reduction strategies	E2

Table 5.3c The Sustainable Development Goal 13 and the Sendai Framework interlinks.

SDG indicators	Goal 13. Take urgent action to combat climate change and its impacts	SFDRR indicators
13.1.1	Number of deaths, missing persons and directly affected persons attributed to disasters per 100,000 population	A1 and B1
13.1.2	Number of countries that adopt and implement national disaster risk reduction strategies in line with the Sendai Framework for Disaster Risk Reduction 2015–2030	E1
13.1.3	Proportion of local governments that adopt and implement local disaster risk reduction strategies in line with national disaster risk reduction strategies	E2

As mentioned above, one of the main features of the SFDRR, compared to its predecessor and several other international frameworks, is the role that the scientific community played in informing its structure. This can be quoted as a compelling case where the interface between science and policy was effectively translated into action.

The SFDRR highlights the role of the Scientific and Technical Advisory Group (STAG[1]), a committee of academic experts coordinated by UNDRR, which actively backs up the global implementation of the framework. The STAG includes academic and scientific experts in disaster and health as well as climate change and health who can actively contribute to a sound policy approach to the implementation of national, regional and international investments in disaster and health.

5.3 The Paris Agreement and the SDGS: Tackling Climate Change and Health

Since the adoption of the Paris Agreement (UNFCCC 2015), scientific communities and international organisations are looking at the impacts that climate change has (and will potentially have) on the health and well-being of individuals and communities (see also Chapter 6).

The topic of health and climate change is often expanded to the concept of human well-being, which is associated with equality, life satisfaction and inclusivity (Marmot 2018). There are several determinants to human well-being, such as education, livelihoods, and safe and clean environment, among others. These factors are vulnerable to the current and future impacts of a changing climate. The consequence is that an appropriate inclusion of a health and well-being consideration in investments that tackle climate change results in a triple win outcome: increased disaster resilience, progresses towards achieving the SDGs and mitigation of negative climate impacts (Paris Agreement).

The impacts of climate change on health have been described in Chapter 3. They include both direct and indirect impacts. Direct impacts encompass physiological effects of exposure to higher temperatures, increasing incidences (or death) due to hazards such as droughts, floods, heatwaves, storms and wildfires (Frumkin and Haines, 2019). WHO estimated that approximately 250,000 deaths annually could occur between 2030 and 2050 due to climate change, given the increase in deadly heatwaves, diarrheal diseases, malaria, dengue, coastal flooding and childhood stunting (WHO 2018).

The indirect impacts of climate change on health are difficult to predict and quantify. To date, this could probably be one of the main areas where further applied research is needed. Climate change produces long-term effects on ecosystems and ecosystem services. The disequilibrium of an ecosystem due to altered climate and weather patterns can have major impacts on food systems and water quality/availability. This implies the risk of food and water insecurity and nutrients deficiencies, especially for those least developing countries and small island developing states which have high climate vulnerabilities and low coping capacity.

Another impact difficult to predict is how (and how fast) the spread of diseases can be driven by ecosystem changes, with exotic vectors migrating and adapting in different geographical areas.

Figure 5.1 showcases the different pathways that climate change can have on health.

Climate change mitigation projects, those which intend to reduce and avoid greenhouse gasses emissions by investing in sectors such as clean energy and sustainable mobility, can have substantial impacts on human health and well-being. Projects which intend to shape the city of the future, the greening of urban settlements, should

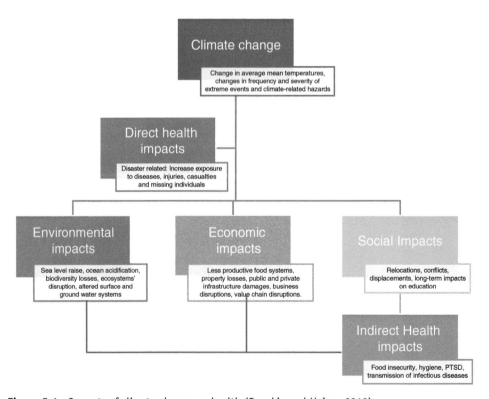

Figure 5.1   Impacts of climate change on health (Frumkin and Haines 2019).

account for the economic and social benefit that this can produce in terms of improved health of the population.

This could result in a major benefit of the investment that, when monetised by an economic analysis, can provide important insight to public decision-makers. For example, a well-planned sustainable mobility intervention, besides reducing and displacing greenhouse gasses, can significantly contribute to the avoidance of fine particles and other air pollutants, improving air quality and reduce the incidence of several chronic and mortal diseases.

Even more, projects in the domain of climate change adaptation can have a substantial impact on health outcomes. An area of investment where key considerations concerning health are of paramount importance are multi-hazard early warning systems. A sound national climate and health strategy should recognise infectious diseases and pandemics as an integral part of what is needed to build the resilience of society and productive systems. National health surveillance systems should be integrated with the existing early warning systems. Monitoring, communicating and preparing for health risks caused by climate-related hazards (such as heatwaves and floods) is pivotal to build climate-resilient societies.

Yet, despite this major primary benefit, health is generally approached as a crosscutting theme in climate investments, rather than an area or a specific sector of intervention, and donors, national institutions and implementing agencies lack specialised trained personnel in health and climate.

In relation to the connection between the SDGs and the Paris Agreement, it is relevant to mention the work of the German Development Institute (DIE) and Stockholm Environmental Institute (SEI) that analyse the linkages between each SDG and the climate change agenda (DIE–SEI 2019). Amongst the 17 SDGs, one targets specifically climate action (SDG 13) and another health and well-being (SDG3) (Dzebo 2017). However, concerning specifically the links between climate change and health, they seem still weak in terms of actual climate national policies and strategies. According to DIE and SEI, the links between SDG 3 and most of NDCs are scarce, with only just over 3% of all NDCs mentioning health.

5.4   Comparative Analysis of the Three Frameworks

So far, the elements in development (SDGs), DRR (Sendai Framework) and climate change (Paris Agreement) have been discussed and international agendas point out a shared vision of how health can be meaningfully embedded in policies and investments at all levels. The analysis leads to two main findings:

1) As discussed above, health and well-being are well integrated and recognised as having a central role in the current development agenda set by the SDGs and in the international DRR agenda framed by the SFDRR. In particular, after ten years of implementation of the HFA, there was a more relevant role of the scientific community in the definition of DRR national and international policies and this contributed significantly to step up the role that health and well-being should have when implementing DRR projects and programmes.

2) While the connection between health and climate change is clear and specialised agencies such as the WHO have well pointed out the areas of intervention to adapt the health sector to a changing climate, the practical inclusion of health consider- ations into climate projects and programmes needs to be better defined. At best, health is considered a co-benefit of climate change projects which addressed adaptation outcomes (e.g., agriculture, food security, water, ecosystems, etc.) or mit- igation results (e.g., air pollution).

As happened in the case of the implementation and review of the HFA, the implementation and review of the Paris Agreement, coupled with the active involve- ment of the scientific community, can help to understand how to build national and international programmes that put human health and well-being at the centre of climate finance.

5.5   Conclusions

Creating an effective science–policy interface during the implementation of the HFA has been a key factor that ensured that the HFA's successor, the SFDRR, fully embedded health as a cornerstone of any DRR interventions at the international and national level. This can be a model for the international and national policies and strat- egies that intend to effectively address climate change adaptation and, to a certain extent, mitigation.

Moreover, this chapter highlights the health nexus amongst the main interna- tional frameworks which influence policies and investments in development, DRR and climate change. It is observed that while the SFDRR and the SDGs have integrated the health nexus better in their respective agendas (compared to their predecessors, the HFA and the Millenium Development Goals (UN 2000)), the role of health in climate policies can be better connected to the actual investments on the ground.

This can be achieved through enhanced exchange amongst the health and the cli- mate communities of practice. International fora and national platforms conveying cli- mate and health practitioners and academics can influence what countries include in their NDCs (and future revisions/updates) to ensure that health is well featured as a key area for climate investments and international finance, rather than a secondary co- benefit of climate change interventions.

Chapter 6 gives further insight into preparedness and response requirements in rela- tion to health and health protection in view of climate change. It also links to the Sendai Framework and the SDGs but also refers to the International Health Regulations (WHO 2016) and the WHO Health and Emergency and Disaster Risk Management Framework of 2019 (WHO 2019).

Note

1   https://www.unisdr.org/partners/academia-research.

References

Aitsi-Selmi, A. and Hiroyuki, S. (2015). The Sendai framework for disaster risk reduction: Renewing the global commitment to people's resilience, health, and well-being. *International Journal of Disaster Risk Science* 6: 164–176.

DIE–SEI. (2019, September 10). https://klimalog.die-gdi.de/ndc-sdg/sdg/3. Retrieved from klimalog.die-gdi.de:https://klimalog.die-gdi.de

Dzebo, A.C. (2017). *Exploring Connections Between the Paris Agreement and the 2030 Agenda for Sustainable Development*. Stockholm: SEI Policy Brief.

Frumkin, H. and Haines, A. (2019, January 11). Global environmental change and noncommunicable disease risks. *Annual Review of Public Health* 40: 261–282.

Innocenti. (2015). When science meets policy: Enhancing governance and management of disaster risks. In: *Hydrometeorological Hazards: Interfacing Science and Policy* (P. Quevauviller), Chapter 2.1. Chichester, UK: John Wiley & Sons, Ltd.

Kelman, I. (2015). Climate change and the Sendai framework for disaster risk reduction. *International Journal of Disaster Risk Science* 6: 117.

Marmot, M. and Bell, R. (2018). The sustainable development goals and health equity. *Epidemiology* 29 (1): 5–7.

UNFCCC. (2015). *Paris Agreement*. Bonn: United Nations.

United Nations. (2000). *United Nations Millennium Declaration*. New York: UN General Assembly.

United Nations. (2015a). *Sendai Framework for Disaster Risk Reduction 2015–2030*. New York: United Nations.

United Nations. (2015b). *Transforming Our World: The 2030 Agenda for Sustainable Development*. New York: UN General Assembly.

WHO. (2018). *COP24 Special Report: Health & Climate Change*. Geneva: WHO.

6

Preparedness and Response in View of Climate Change Impacting on Health Challenges

Virginia Murray[1] and Lidia Mayner[2]

[1]Head of Global Disaster Risk Reduction and a COVID-19 senior public health advisor
Public Health England

[2]Torrens Resilience Institute, College of Nursing and Health Sciences, Flinders University

6.1 Introduction

At the time of writing, the COVID-19 pandemic is a timely reminder of how hazards within the complex and changing global risk landscape can affect lives, livelihoods and health. It is increasingly recognised that climate change will be a more complex phenomena to address when combined to other phenomena such as the COVID-19 pandemic having to cope with the combined effects which will impact on mental health and associated emergencies (Marazziti et al. 2021).

Created by the World Meteorological Organisation (WMO) and United Nations Environment Programme in 1988, the Inter-governmental Panel on Climate Change (IPCC) published an assessment report every five to six years after reviewing and assessing the most recent scientific, technical and socio-economic information relevant to the understanding of climate change and related disasters. Health impacts of climate disasters have been gaining coverage in the assessment reports since the mid-1990s. Of note, the IPCC report on Managing the Risks of Extreme Events and Disasters to Advance Climate Change Adaptation (IPCC 2012) brought attention globally to the impact of extreme events and disasters. A series of case studies were used to highlight these issues (Murray et al. 2012). In the 2014 IPCC report, it was identified that climate change would exacerbate existing health problems and increase the number of ill

Hydrometeorological Extreme Events and Public Health, First Edition. Edited by
Franziska Matthies-Wiesler and Philippe Quevauviller.
© 2022 John Wiley & Sons Ltd. Published 2022 by John Wiley & Sons Ltd.

people in many countries, especially in developing countries, which would threaten sustainable development (IPCC 2014).

As defined by IPCC, climate change refers to 'a change in the state of the climate that can be identified (e.g., by using statistical tests) by changes in the mean and/or the variability of its properties, and that persists for an extended period, typically decades or longer' (IPCC 2018). From wider reporting, it is well recognised that hydrometeorological extreme incidents and emergencies can have significant effects on people's health, including loss of life (see Chapter 5).

Climate change is a driver for the increasing number of hydrometeorological extreme incidents and emergencies (see Chapter 2). The World Health Organisation (WHO) reports that 'climate change is impacting human lives and health in a variety of ways. It threatens the essential ingredients of good health – clean air, safe drinking water, nutritious food supply, and safe shelter – and has the potential to undermine decades of progress in global health.'WHO considers that 'Between 2030 and 2050, climate change is expected to cause approximately 250,000 additional deaths per year, from malnutrition, malaria, diarrhoea and heat stress alone. The direct damage costs to health is estimated to be between USD 2–4 billion per year by 2030.' (WHO 2021).

To try and address these and other issues, the United Nations (UN) member states adopted three landmark agreements in 2015: the Sendai Framework for Disaster Risk Reduction 2015–2030 ('the Sendai Framework') (UNDRR 2015), the Paris Agreement on Climate Change (UN 2015a) and the Sustainable Development Goals (UN 2015b).

6.2 The Sendai Framework for Disaster Risk Reduction 2015–2030

The Sendai Framework is a global UN strategy for addressing disaster risk and resilience and aims for the following outcome by 2030: 'The substantial reduction of disaster risk and losses in lives, livelihoods and health and in the economic, physical, social, cultural and environmental assets of persons, businesses, communities and countries' (UNDRR 2015). The framework has thirteen guiding principles, seven global targets and four priority areas for action which are:

Priority 1: Understanding disaster risk
Priority 2: Strengthening disaster risk governance to manage disaster risk
Priority 3: Investing in disaster risk reduction for resilience
Priority 4: Enhancing disaster preparedness for effective response and to 'Build Back Better' in recovery, rehabilitation and reconstruction

It adopts an all hazards approach to managing disaster risk arising from environmental, technological and biological hazards and risks (UNDRR 2015). For the first time in any of the disaster risk reduction UN agreement, the Sendai Framework included a list of seven global targets (Box 6.1). These, with the 38 indicators, require biannual reporting from all UN member states.

New threats reveal new challenges for managing the health risks and effects of emergencies. Deaths, injuries, diseases, disabilities, psychosocial problems and other health impacts can be avoided or reduced by skilled disaster management involving health

---

Box 6.1   The Sendai Framework for Disaster Risk Reduction 2015–2030 global targets (UNDRR 2015) (paragraph 18)

a) The Sendai Framework for Disaster Risk Reduction 2015–2030 global targets (UNDRR 2015) (paragraph 18)
b) Substantially reduce global disaster mortality by 2030, aiming to lower average per 100,000 global mortality rate in the decade 2020–2030 compared to the period 2005–2015.
c) Substantially reduce the number of affected people globally by 2030, aiming to lower average global figure per 100,000 in the decade 2020–2030 compared to the period 2005–2015.
d) Reduce direct disaster economic loss in relation to global gross domestic product (GDP) by 2030.
e) Substantially reduce disaster damage to critical infrastructure and disruption of basic services, among them health and educational facilities, including through developing their resilience by 2030.
f) Substantially increase the number of countries with national and local disaster risk reduction strategies by 2020.
g) Substantially enhance international cooperation to developing countries through adequate and sustainable support to complement their national actions for implementation of this Framework by 2030.
h) Substantially increase the availability of and access to multi-hazard early warning systems and disaster risk information and assessments to the people by 2030.

---

and other sectors. The Sendai Framework places strong emphasis on resilient health systems at primary, secondary and territorial levels, and by advocating for developing the capacity of health workers in understanding disaster risk and applying disaster risk approaches in their work (UNDRR 2015).

The linkage between climate change and disaster risk is the most apparent through natural disasters exacerbated by climate change (see Chapter 2). As it is inducing and exacerbating related natural disasters, e.g., flooding, storm and drought, climate change is mentioned 15 times and considered a major driver of disaster risk in the Sendai Framework, where countries are urged to incorporate climate change scenarios in their disaster risk assessments (Kelman 2015, 2017).

6.2.1   Paris Agreement on Climate Change and Emergency Preparedness

The Paris Agreement is a legally binding international treaty on climate change. It was adopted by 196 Parties at the Conference of the Parties (COP) 21 in Paris, on 12 December 2015 and entered into force on 4 November 2016 (UN 2015a). Its goal is to limit global warming to well below 2 °C, preferably to 1.5 °C, compared to pre-industrial levels. The Paris Agreement is a landmark in the multilateral climate change process because, for the first time, a binding agreement brings all nations into a common cause to undertake ambitious efforts to combat climate change and adapt to its effects.

In article 8 of the Paris Agreementit recognises that:

Acknowledging that climate change is a common concern of humankind, Parties should, when taking action to address climate change, respect, promote and consider their respective obligations on human rights, the right to health, the rights of indigenous peoples, local communities, migrants, children, persons with disabilities and people in vulnerable situations and the right to development, as well as gender equality, empowerment of women and intergenerational equity.

In article 4, it documents that 'areas of cooperation and facilitation to enhance understanding, action and support may include:

(a) Early warning systems;
(b) Emergency preparedness;
(c) Slow onset events;
(d) Events that may involve irreversible and permanent loss and damage;
(e) Comprehensive risk assessment and management;
(f) Risk insurance facilities, climate risk pooling and other insurance solutions;
(g) Non-economic losses; and
(h) Resilience of communities, livelihoods and ecosystems.' (UN 2015a)

In recognition of the need for enhanced understanding, action and support to include emergency preparedness, the WHO is encouraging UN member states to follow and sustain this agreement where 'health resilience is strongly promoted throughout' (UNDRR 2015).

6.3 Sustainable Development Goals

The 2030 Agenda for Sustainable Development (UN 2015c) and the related sustainable development goals (SDGs) adopted in September 2015 at the United Nations Sustainable Development Summit (UN 2015d) expanded beyond the eight original goals of its predecessor the Millennium Development Goals (MDGs) (UN 2000). The 17 SDGs with 169 targets seek to address and balance the three dimensions of sustainable development: economic growth, social inclusion and environmental protection. These goals are designed to be integrated and indivisible, aiming to enable risk-resilient and sustainable development for all countries (UN 2015b).

As to the 2030 Agenda for Sustainable Development, Goal 13 commits all countries to take urgent action to combat climate change and its impacts, while Goal 3 calls for the strengthening of national capacity for early warning, risk reduction and management of national and global health risks, which are closely linked to climate change-related disasters. The relevance of climate change and health also underpins many other goals, such as Goal 2 on improving food security and nutrition (given that extreme temperatures and rainfall resulting from climate change could lead to crop failure and food shortage) and Goal 6 on water and sanitation (given that climate change-enhanced disasters like flooding and sea-level rise could result in water contamination and salination). The interlinkage of these goals with the Sendai Framework are summarised in Figure 6.1 (WHO 2020).

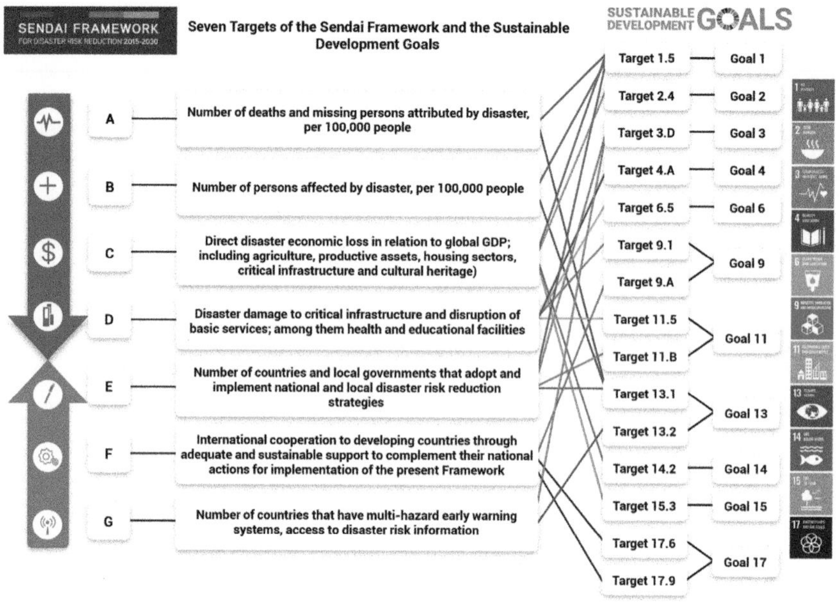

Figure 6.1   Links between Sendai Framework targets and Sustainable Development goals (WHO 2020).

While emergencies affect everyone, they disproportionately affect those who are the most vulnerable. The needs and rights of the poorest, as well as women, children, people with disabilities, older persons, migrants, refugees and displaced persons, and people with chronic diseases must be at the centre.

This chapter considers the tools for public health risk management in relation to hydrological extreme events and what preparedness and response processes are needed in view of climate change for these hazards. It addresses what are the hydrometeorological hazards; shares summaries of the WHO International Health Regulations (WHO 2016) and the WHO Health Emergency and Disaster Risk Management (Health EDRM) Framework (WHO 2019), which emphasises the critical importance of prevention, preparedness and readiness, together with response and recovery, to save lives and protect health. Case studies are used to illustrate the principles described.

6.4   What are Hydrometeorological Extreme Events and How are They Defined?

Defining hazards has always been complex. The World Meteorological Organisation is mandated to provide the framework for such international cooperation on weather, climate and the water cycle, which know no national boundaries, requiring international cooperation at a global scale to be essential for the development of meteorology, climatology and operational hydrology as well as to reap the benefits from their application (World Meteorological Organisation 2020).

Hazard information when combined with exposure, vulnerability and capacity is fundamental to all aspects of disaster risk management, for example, for risk assessments (before, during and after events); policy development and review; planning and implementation of risk management measures; monitoring and reporting; and for documenting the losses and damage from hazardous events including disasters (World Meteorological Organisation 2014).

In 2015, at the Seventeenth Session of the World Climate Congress (Cg-17), it was decided to standardise hazard and extreme event information, including the creation or adoption of a system of assigning a unique identifier to each event so that events can be catalogued and linked to data on associated damages and losses (World Meteorological Organisation 2019). The World Meteorological Organisation has approved an initial list of hazards to be used with its cataloguing initiative. This comprises hazards under the World Meteorological Organisation mandate for which their names and definitions have been agreed by its members (countries).

Now World Meteorological Organisation is working on the standardisation of data and meta-data for more than 20 meteorological, hydrological and climate-related hazards for enhanced disaster risk reduction, as well as geo-referencing the loss and damage data as one of its key aspects. The World Meteorological Organisation has approved a methodology (World Meteorological Organisation 2019) that will provide national partners with the ability to more systematically and accurately attribute risks and impacts (loss and damage) of hazards to causal phenomena (hazards). The methodology centres on an event-unique identifier that includes parameters that detail the event hazard name, temporal and spatial information, and linkage to other events as well as specific contextual information that will aid stakeholders in loss and damage accounting in attributing losses to the causal phenomena. The two central parameters of this methodology are the hazard name and linkages.

For the standardisation work by the World Meteorological Organisation, data has been fed into the work initiated formally in May 2019 by the UN Office for Disaster Risk Reduction (UNDRR) and the International Science Council (ISC). Both offices jointly established a technical working group to identify the full scope of the 'all hazards' relevant to the Sendai Framework as a basis for countries to review and strengthen their risk reduction policies and operational risk management practices.

As a scientific undertaking, the technical working group was guided by the definition of 'hazard' adopted by the United Nations General Assembly (UNGA) in February 2017; namely, 'a process, phenomenon or human activity that may cause loss of life, injury or other health impacts, property damage, social and economic disruption or environmental degradation.' This definition covers a broader scope of hazards than has traditionally been the case in the field of disaster risk reduction and expands the definition of hazard to include processes and activities. The hazard list comprises 302 hazards grouped according to 8 clusters: meteorological and hydrological hazards, extraterrestrial hazards, geohazards, environmental hazards, chemical hazards, biological hazards, technological hazards and societal hazards. Although this hazard list is considered to be the most useful one at the present time, it is not a definitive list and needs regular review and updating.

From the UNDRR/ISC Hazard Definition and Classification Review (2020), the following 57 hazards have been identified as the principle hydrometeorological extreme events. By hazard cluster type, the principle hydrometeorological hazards include::

- **Convective-related** Downburst; Lightning (Electrical storm); Thunderstorm
- **Flood** Coastal flood; Estuarine flood; Flash flood; Fluvial (riverine flood); Groundwater flood; Ice-jam flood including debris; Ponding flood; Snowmelt flood; Surface water flooding; Glacial lake outburst flood
- **Lithometeors** Black carbon (Brown clouds); Dust storm or Sandstorm; Fog; Haze; Polluted air; Sand haze; Smoke
- **Marine:** Ocean acidification; Sea-water intrusion; Sea ice (ice bergs); Ice flow; Storm surge; Storm tides; Tsunami
- **Precipitation-related** Blizzard; Drought; Hail; Ice storm; Snow/Ice; Snow storm; Sub-tropical cyclone
- **Pressure-related** Depression or cyclone (low pressure area); Extra-tropical cyclone
- **Temperature-related** Cold wave; Freeze; Frost (hoar frost); Freezing rain; Glaze; Ground frost; Heatwave; Icing (including ice); Thaw
- **Terrestrial** Avalanche
- **Wind-related** Gale (strong gale); Subtropical storm; Tropical cyclone (cyclonic wind, rain [storm] surge); Tropical storm; Tornado; Wind

6.5 Public Health Risk Management in Relation to Hydrological Extreme Events

With the increasing drivers of climate change, unplanned urbanisation, population growth and displacement, antimicrobial resistance and state fragility contributing to the increasing frequency, severity and impacts of many types of hazardous events, these may lead to emergencies and disasters without effective public health risk management. COVID-19 has shown that the health, economic, political and societal consequences of such events can be devastating. As a result, all communities are at risk of emergencies and disasters including those associated with infectious disease outbreaks, conflicts, and including hydrometeorological, technological and other hazards.

6.5.1 International Health Regulations 2005

The International Health Regulations (IHR), agreed by the World Health Assembly in 2005 and revised in 2007 (WHO 2016), were introduced to protect the world from global health threats. The need for such an internationally binding instrument was identified following the emergence of severe acute respiratory syndrome (SARS) in 2003, which demonstrated how quickly a new disease could spread. As a result, 196 countries acknowledged a responsibility for collective action against infectious diseases, and food and environmental contamination, deemed to be public health emergencies of international concern. This binding instrument of international law entered into force on 15 June 2007. The stated purpose and scope of the IHR are:

> *to prevent, protect against, control and provide a public health response to the international spread of disease in ways that are commensurate with and restricted to public health risks, and which avoid unnecessary interference with international traffic and trade.*

Member states are responsible for developing, strengthening and maintaining the capacity to detect, assess the risks of, and respond to public health emergencies. Furthermore, countries are required to notify the WHO of these threats through their designated National IHR Focal Point within 24 hours. The regulations mandate countries to assess their capacities for disease surveillance and response and report whether these are sufficient to meet their obligations.

WHO also provides support to countries in strengthening and maintaining their capacities for ensuring rapid detection, verification and response to public health risks, as it develops and provides tools, guidance and training. In this way the support from WHO focuses on the priority needs identified by the WHO Regional and Country Offices, in order to help each country to meet its IHR commitment.

While countries have strengthened capacities to reduce the health risks and consequences of emergencies and disasters through the implementation of multi-hazard disaster risk management, the IHR (WHO 2016) and health system strengthening, many communities remain highly vulnerable to a wide range of hazardous events and may not be using the IHR to preparedness and response in view of climate change.

Fragmented approaches to different types of hazards, including those driven by climate change, have resulted in an over-emphasis on reacting to, instead of preventing events and preparing properly to be ready for response, and gaps in coordination across the entire health system, and between health and other sectors. This has hindered the ability of communities and countries to achieve optimal development outcomes, including for public health. Reducing the health risks and consequences of emergencies including those driven by climate change is vital to local, national and global health security and to build the resilience of communities, countries and health systems.

6.5.2 WHO Health Emergency and Disaster Risk Management Framework 2019

Learning from many biological and other emergencies and disasters, it was identified that in order to address current and emerging risks to public health and the need for effective utilisation and management of resources, the conceptual frame or paradigm of 'health emergency and disaster risk management' has been developed to consolidate contemporary approaches and practice. The WHO Health Emergency and Disaster Risk Management Framework (WHO Health EDRM Framework) is fully consistent with and helps to align policies and action for health security, disaster risk reduction, humanitarian reform, and sustainable development agendas, and most importantly for climate change (WHO 2019).

In summary, the WHO Health EDRM Framework:

- provides a common language and a comprehensive approach that can be adapted and applied by all actors in health and other sectors who are working to reduce health risks and consequences of emergencies and disasters
- focuses on improving health outcomes and well-being for communities at risk in different contexts, including in fragile, low- and high-resource settings
- emphasises assessing, communicating and reducing risks across the continuum of prevention, preparedness, readiness, response and recovery, and building the resilience of communities, countries and health systems.
- is derived from the disciplines of risk management, emergency management, epidemic preparedness and response, and health systems strengthening.
- is fully consistent with and helps to align policies and actions for health security, disaster risk reduction, humanitarian action, climate change and sustainable development.

The expected outcome of WHO Health Emergency and Disaster Risk Management Framework (WHO 2019) is that 'countries and communities have stronger capacities and systems across health and other sectors resulting in the reduction of the health risks and consequences associated with all types of emergencies and disasters'.

The WHO Health EDRM Framework is founded on the following set of core principles and approaches that guide policy and practice and include a risk-based approach; comprehensive emergency management (across prevention, preparedness, readiness, response and recovery); the all-hazards approach including hydrometeorological extreme events where climate change is a driver; an inclusive, people- and community-centred approach; a multisectoral and multidisciplinary collaboration; 'whole-of-health system-based'; and is based on ethical considerations.

The WHO Health EDRM Framework comprises a set of functions and components that are drawn from multisectoral emergency and disaster management, capacities for implementing the IHR (WHO 2016), health system building blocks, and good practices from regions, countries and communities (Box 6.2). The Framework focuses mainly on the health sector, noting the need for collaboration with many other sectors that make substantial contributions to reducing health risks and consequences.

Box 6.2 WHO Health EDRM Framework (2019) functions are organised under the following components

- **Policies, Strategies and Legislation:**Defines the structures, roles and responsibilities of governments and other actors for Health EDRM; includes strategies for strengthening Health EDRM capacities.
- **Planning and Coordination:**Emphasises effective coordination mechanisms for planning and operations for Health EDRM.
- **Human Resources:** Includes planning for staffing, education and training across the spectrum of Health EDRM capacities at all levels, and the occupational health and safety of personnel.
- **Financial Resources:**Supports implementation of Health EDRM activities, capacity development and contingency funding for emergency response and recovery.

- **Information and Knowledge Management**:Includes risk assessment, surveillance, early warning, information management, technical guidance and research.
- **Risk Communications**:Recognises that communicating effectively is critical for health and other sectors, government authorities, the media and the general public.
- **Health Infrastructure and Logistics**: Focuses on safe, sustainable, secure and prepared health facilities, critical infrastructure (e.g., water, power), and logistics and supply systems to support Health EDRM.
- **Health and Related Services**:Recognises the wide range of healthcare services and related measures for Health EDRM.
- **Community Capacities for Health EDRM**:Focuses on strengthening local health workforce capacities and inclusive community-centred planning and action.
- **Monitoring and Evaluation**:Includes processes to monitor progress towards meeting Health EDRM objectives, including monitoring risks and capacities and evaluating the implementation of strategies, related programmes and activities.

The WHO Health EDRM Framework provides three useful annexes giving the WHO classification of hazards; a helpful checklist on the components and functions of health emergency and disaster risk management; and a list of stakeholder groups for health emergency and disaster risk management (WHO 2019).

Climate change is a cross-cutting global issue with significant impact on disaster risk management and wider sustainable development, which explains its prominence in the related international risk-resilient and sustainable development agendas. Since health is an essential component of climate-related disasters management, the WHO Health EDRM Framework provides a comprehensive tool to synergise these global efforts by means of improving health outcomes in disasters in particular. With the far-reaching impact of climate change over different aspects of human living and development, climate change adaptation now goes hand-in-hand with mitigation, which was traditionally the major focus of climate change policy. The WHO Health EDRM Framework in the climate–disaster context is to call for a wide range of actions to improve the health outcomes, in particular the building of health systems and community resilience and the development of effective early warning systems.

6.6   Public Health Risk Management – Some Specific Hydrometeorological Events Issues

Public health risk management has identified specific hydrometeorological extreme events issues, two of which are addressed below, namely infectious diseases and mental health.

6.6.1   Infectious Disease Associated with Hydrometeorological Extreme Events

A recent United Nations Environmental Programme report acknowledged that many infectious diseases are highly sensitive to climate variability and change (Cissé et al. 2018). This summary below addresses findings related to flooding and drought events.

6.6.1.1 Flood-Related Infectious Diseases The links between floods and infectious diseases have been introduced in Chapter 4. Some studies showed that the frequency of infectious diseases can increase in the weeks to months after flooding. Water-borne diseases occur acutely within 7 days, rodent-borne diseases are more frequently associated as occurring within 1–4 weeks and vector-borne diseases are likely to be move long term, often occurring more than 4 weeks after a flood event (Brown and Murray 2013). However, there remains scientific uncertainty about the strength of association between infectious disease incidence and flooding (Du et al. 2010). Floods can cause population displacement and changes in population density, raise concern about waste management and the availability of clean water, as well as affect the availability and access to healthcare services. Overall, the risk of infectious disease following flooding is context-specific, differs between countries, and is dependent upon a number of synergistic factors. Outbreaks of leptospirosis and diarrheal diseases following flooding have been documented in Europe (Desai et al. 2009; Marcheggiani et al. 2010; Reacher et al. 2004), but the evidence of increased incidence of vector-borne diseases following flooding is lacking because the time lag before onset can be several months (Waring and Brown, 2005). Past studies have indicated possible associations between vector-borne diseases and flooding in Europe (Brown et al. 2014; Hubálek and Halouzka 1999; Pellizer et al., 2006; Reacher et al. 2004). In addition, European residents may be exposed to these risks while travelling. Foreign relief workers can potentially introduce infectious diseases into an area affected by flooding and these workers may be susceptible to endemic diseases that are more prevalent because of the flood.

Surveillance and early warning systems in flood-affected areas are fundamental to understanding the impact of flooding on infectious disease incidence. Such data may be less accessible if flood damage to pre-existing public health infrastructure has exacerbated weaknesses in a disease surveillance system. Where available surveillance should provide information to reduce current and future vulnerability and should be able to detect epidemic-prone diseases, assuring access to clean water, proper sanitation, adequate shelter and primary healthcare services is essential. Furthermore, it is difficult to attribute an increase in infectious disease incidence solely to a flood event, and therefore this issue may be under-investigated and under-reported. In Europe, for maintenance and continuous adaptation and improvement of public health, Brown and Murray (2013) recommended that research needs to improve the understanding of the association between infectious diseases and flooding by developing:

- more robust epidemiological studies on infectious diseases covering the pre-, mid- and post-flood periods
- further research assessing the effectiveness of public health interventions minimising risk from infectious diseases following flooding
- investigation of infectious disease incidence following smaller flood events
- analysis of the differences between summer and winter flooding on infectious disease incidence
- analysis of the differences between flash and riverine flooding on infectious disease incidence

6.6.1.2 Drought-Related Infectious Diseases The most recent data available from the international disasters database estimate that more than 50 million people around the world were affected by drought in 2011 (Stanke et al. 2013). Despite their dangerous global impacts, the health effects of droughts are poorly researched and understood, partly due to the complexities involved in ascertaining the beginning and end of the events, individual and population exposures and their accumulated effects over time (UNDRR, GAR Special Report on Driygh, 2021). In addition, many of its effects are indirect (Berman et al. 2017; Ebi and Bowen 2016; Vins et al. 2015) and thus inherently difficult to attribute.

During droughts, paucity of water may increase the risk of infectious disease and these infections can be associated with water-related disease caused by faecal/urine pollution, water-related disease caused by poor handwashing/hygiene, and skin, eye and louse-borne diseases that occur when there is lack of water for personal hygiene (Stanke et al. 2013).

Water-related disease associated with contamination by faeces or urine has been found; for example, in the case of an increased number of animals being drawn to drink by a river or borehole or from poor-quality irrigation water (Shehane et al. 2005; Lake and Barker 2018). Infected humans can also contribute to transmission around a water source, because the more users there are, the more are at risk and the more likely the occurrence of an outbreak of infectious disease. Associated risks might be further exacerbated if the water source in question is running at low levels, pathogens are more concentrated than normal, and it is more likely that anybody consuming the water would ingest a minimum infective dose. Diseases which are transmitted by water and hence potentially affected by drought include amoebiasis, hepatitis A, salmonellosis, schistosomiasis, shigellosis, typhoid and paratyphoid (enteric fever). Evidence is, however, scarce. Few papers directly addressed this topic and were limited to drought-associated cholera and leptospirosis (Levy et al. 2016). In all these studies, it is important to note that drought was not the only exposure underlying the outbreak – a whole chain of risk factors was responsible.

Though it did not formally explore disease mechanisms, a 1978 study from Wales supports the plausible hypothesis that poor hand-washing (due to decreased water supply during drought) can also contribute to drought-associated diarrhoeal disease (Burr et al. 1978). Areas with the longest cuts to mains supply had significantly greater diarrhoea prevalence than areas with shorter-lasting or no daily cuts. Mains water meant that source contamination is an unlikely explanation for these results (Burr et al. 1978). Furthermore, eye infections including conjunctivitis were associated with drought-related lack of water for washing but skin infections associated with lack of water for washing include scabies and impetigo were not so clearly related (Thacker et al. 1980).

6.6.2 Impacts of Climate Change Emergencies on Mental Health

Climate change can have widespread cascading effects brought about by high temperature, drought, wildfires, loss of property, and loss of income which all combine to impact on people affected by such losses and suffering serious consequences on their mental health (Chapters 3 and 4). It is becoming increasingly evident that apart from the climate-related events and disasters, climate change is impacting on mental health. A Scopus search returned 208 publications for the period 2007–2016 for the term climate change and mental health (Berry et al. 2018), while a recent (March 2021) Scopus search yielded 653 publications for the period 2016–2021.

A number of authors have described (e.g., Clayton 2021; Seritan and Seritan 2020) the onset of mental health conditions such as anxiety, depression, post-traumatic stress disorder (PTSD) and suicide increasing over time, following natural disasters and extreme weather events. An example described (Seritan and Seritan 2020) that after Hurricane Katrina (2005) there was a 6% increase in PTSD, suicide ideation an increase of 3.6% and suicide plans an increase of 1.5%. Silveira et al. (2021) looked at the mental health outcomes following the wildfires in California in 2018. They found (statistically significant) that direct exposure to large-scale fires increased the risk of mental health disorders, in particular PTSD and depression. They also found that this increase was also due to the quality of sleep, a finding also shared by Ogunbode et al. (2021), who looked at insomnia and negative climate-related emotions.

Climate change is impacting adversly on the environment and these effects result in feelings of loss from changes made to personally significant places. Usher et al. (2019) described this as ecological grief and with the global climate situation worsening others have used terms such as eco-anxiety, solastalgia or eco-angst (cited in Usher et al. 2019). Eco-anxiety is the stress arising from changes in to the environment and one's knowledge of them (Usher et al. 2019).

Berry et al. (2018) described using a systems approach for addressing the issue of climate change and mental health, whereby systems thinking could encompass the understandings from many different disciplines. Here systems would encompass geo-political, socioeconomic, ecological and environmental factors that impact on mental health – thus expanding on the notion that mental health is person-based but more so community-based.

6.7   Conclusions and Suggested Ways Forward

Climate change is now recognised increasingly as a ubiquitous driver of emergency and disasters. Emergency preparedness to address an all-hazards approach is critical. Health emergency preparedness along with adaptation is most effective when it is integrated into broader national policies and programmes, notably by evaluating and ensuring the health benefits of sectoral adaptation activities in agriculture, water and sanitation, infrastructure, transport, energy and urban planning, as well as by creating synergies with complementary initiatives such as the Sendai Framework and the SDGs, the Paris Agreement and the International Health Regulations, among others. In summary, integration of health into broader policy frameworks is critical.

The following are a selection of issues that require urgent consideration:

- Review of emergency preparedness and the implementation of the WHO Health Emergency and Disaster Risk Framework and the International Health Regulations is needed by all countries. This will include delivery of public health risk management in relation to hydrometeorological extreme events.
- Building effective early warning and monitoring systems. This includes the development and implementation of multi-hazard early warning systems along with effective planning and risk management in line with the global targets of the Sendai Framework and the SDGs. Of note, progress is hindered by several factors, including widely varying surveillance capacity amongst countries and regions, and a lack of clear, standardised definitions of extreme events.

- Investing in capacity-building and preparedness is essential across the health sector and emergency preparedness partnership spectrum and developing health action plans and enhancing health workforce preparedness to climate impacts would deliver immediate and long-lasting results that would improve adaptive capacity in health and strengthen the resilience of health systems.
- Improving basic measures in water, sanitation and hygiene. Development-related measures in water, sanitation and hygiene, as well as in food safety, can prevent many of the additional deaths due to climate sensitive infectious diseases.

References

Berman, J.D., Ebisu, K., Peng, R.D. et al. 2017). Drought and the risk of hospital admissions and mortality in older adults in western USA from 2000 to 2013: A retrospective study. *Lancet Planet Health* 1 (1): e17–e25.

Berry, H.L., Waite, T.D., Dear, K.B.G. et al. (2018). The case for systems thinking about climate change and mental health. *Nature Climate Change* 8 (4): 282–290. doi:10.1038/s41558-018-0102-4.

Brown, L., Medlock, J., and Murray, V. (2014). Impact of drought on vector-borne diseases – How does one manage the risk? *Public Health* 128 (1): 29–37. doi:10.1016/j.puhe.2013.09.006.

Brown, L. and Murray, V. (2013). Examining the relationship between infectious diseases and flooding in Europe A systematic literature review and summary of possible public health interventions. *Disaster Health* 1 (2): 1–11. April/May/June 2013; © 2013 Landes Bioscience.

Burr, M.L., Davis, A.R., and Zbijowski, A.G. (1978). Diarrhoea and the drought. *Public Health* 92 (2): 86–87. doi:10.1016/s0033-3506(78)80034-1.

Cissé, G., Menezes, J.A., and Confalonieri, U. (2018). *Climate Sensitive Infectious Diseases in UNEP 2018. The Adaptation Gap Report 2018*. Nairobi, Kenya: United Nations Environment Programme (UNEP).

Clayton, S. (2021). Climate change and mental health. *Current Environmental Health Reports*. doi:10.1007/s40572-020-00303-3.

Desai, S., Van Treeck, U., Lierz, M. et al. (2009). Resurgence of field fever in a temperate country: An epidemic of leptospirosis among seasonal strawberry harvesters in Germany in 2007. *Clinical Infectious Diseases : An Official Publication of the Infectious Diseases Society of America* 48: 691–697. PMID:19193108. http://dx.doi.org/10.1086/597036.

Du, W., FitzGerald, G.J., Clark, M. et al. (2010). Health impacts of floods. *Prehospital and Disaster Medicine* 25 (3): 265–272.

Ebi, K.L. and Bowen, K. (2016). Extreme event as sources of health vulnerability: Drought as an example. *Weather and Climate Extremes* 11: 95–102.

Hubálek, Z. and Halouzka, J. (1999). West Nile fever–a re-emerging mosquito-borne viral disease in Europe. *Emerging Infectious Diseases* 5 (5): 643–650. doi:.10.3201/eid0505.990505.

IPCC (2012). *Summary for Policymakers. In: Managing the Risks of Extreme Events and Disasters to Advance Climate Change Adaptation* (ed. C.B. Field, V. Barros, T.F. Stocker, et al.), 1–19. A Special Report of Working Groups I and II of the Intergovernmental Panel on Climate Change. Cambridge, UK and New York, USA: Cambridge University Press.

IPCC (2014). *Climate Change 2014: Synthesis Report. Contribution of Working Groups I, II and III to the Fifth Assessment Report of the Intergovernmental Panel on Climate Change.* (Core Writing Team, ed. R.K. Pachauri and L.A. Meyer), 151 pp. Geneva, Switzerland: IPCC. https://archive.ipcc.ch/pdf/special-reports/srex/SREX_FD_SPM_final.pdf.

IPCC (2018). Annex I: Glossary [Matthews, J.B.R. (ed.)]. In: *Global Warming of 1.5°C. An IPCC Special Report on the Impacts of Global Warming of 1.5 °C above Pre-industrial Levels and Related Global Greenhouse Gas Emission Pathways, in the Context of Strengthening the*

Global Response to the Threat of Climate Change, Sustainable Development, and Efforts to Eradicate Poverty (ed. V. Masson-Delmotte, P. Zhai, H.-O. Pörtner, et al.).

Kelman, I. (2015). Climate change and the Sendai framework for disaster risk reduction. *International Journal of Disaster Risk Science* 6: 117–127. doi:.10.1007/s13753-015-0046-5.

Kelman, I. (2017). Linking disaster risk reduction, climate change, and the sustainable development goals. *Disaster Prevention and Management* 26 (3): 254–258. doi:10.1108/DPM-02-2017-0043.

Lake, I.R. and Barker, G.C. (2018). Climate change, foodborne pathogens and illness in higher-income countries. *Current Environmental Health Reports* 5 (1): 187–196.

Levy, K., Woster, A.P., Goldstein, R.S. et al. (2016). Untangling the impacts of climate change on waterborne diseases: A systematic review of relationships between diarrheal diseases and temperature, rainfall, flooding, and drought. *Environmental Science & Technology* 50 (10): 4905–4922. https://pubmed.ncbi.nlm.nih.gov/27058059. Accessed on 13 March 2021.

Marazziti, D., Cianconi, P., Mucci, F. et al. (2021). Climate change, environment pollution, COVID-19 pandemic and mental health. *Science of the Total Environment* 773. doi:.10.1016/j.scitotenv.2021.145182.

Marcheggiani, S., Puccinelli, C., Della Bella, V. et al. (2010). Risks of water-borne disease outbreaks after extreme events. *Toxicological and Environmental Chemistry* 92: 593–599. http://dx.doi.org/10.1080/02772240903252140.

Murray, V., McBean, G., Bhatt, M. et al. (2012). Case studies. In: *Managing the Risks of Extreme Events and Disasters to Advance Climate Change Adaptation* (ed. C.B. Field, V. Barros, T.F. Stocker, et al.), 487–542. A Special Report of Working Groups I and II of the Intergovernmental Panel on Climate Change (IPCC). Cambridge, UK and New York, USA: Cambridge University Press.

Ogunbode, C.A., Pallesen, S., Böhm, G. et al. (2021). Negative emotions about climate change are related to insomnia symptoms and mental health: Cross-sectional evidence from 25 countries. *Current Psychology*. doi: 10.1007/s12144-021-01385-4.

Pellizzer, P., Todescato, A., Benedetti, P. et al. (2006). Leptospirosis following a flood in the Veneto area, North-east Italy. *Annali Di Igiene: Medicina Preventiva E Di Comunita* 18: 453–456. PMID: 17089960.

Reacher, M., McKenzie, K., Lane, C, et al. (2004). Lewes flood action recovery team: Health impacts of flooding in Lewes: A comparison of reported gastrointestinal and other illness and mental health in flooded and non-flooded households. *Communicable Disease and Public Health/PHLS* 7 (1): 56–63.

Seritan, A.L. and Seritan, I. (2020). The time is now: Climate change and mental health. *Academic Psychiatry* 44 (3): 373–374. doi: 10.1007/s40596-020-01212-1.

Shehane, S.D., Harwood, V.J., Whitlock, J.E et al. (2005). The influence of rainfall on the incidence of microbial faecal indicators and the dominant sources of faecal pollution in a Florida river. *Journal of Applied Microbiology* 98 (5): 1127–1136.

Silveira, S., Kornbluh, M., Withers, M.C. et al. (2021). Chronic mental health sequelae of climate change extremes: A case study of the deadliest californian wildfire. *International Journal of Environmental Research and Public Health* 18 (4): 1–15. doi: 10.3390/ijerph18041487.

Stanke, C., Kerac, M.,., Prudhomme, C. et al. (2013). Health effects of drought: A systematic review of the evidence. *PLOS Currents Disasters*. June 5. Edition 1. doi:10.1371/currents.dis.7 a2cee9e980f91ad7697b570bcc4b004.

Thacker, S.B., Music, S.I., Pollard, R.A. et al. (1980). Acute water shortage and health problems in Haiti. *Lancet* 1 (8166): 471–473.

UNDRR (2015). United Nations office for Disaster Risk Reduction, Sendai framework for disaster risk reduction 2015–2030. Available at: https://www.preventionweb.net/files/43291_sendaiframeworkfordrren.pdf. Accessed on 26 September 2020

UNDRR/ISC (2020). UNDRR/ISC Sendai Hazard Definition and Classification Review Technical Report. Available at Hazard definition and classification review | UNDRR. Accessed on 1 December 2020.

UNDRR (2021). Global Assessment Report on Disaster Risk Reduction (GAR) Special Report on Drought 2021. Geneva. ISBN: 9789212320274. GenevaAvailable at: https://www.undrr.org/publication/gar-special-report-drought-202. Accessed on 5 September 2021.

United Nations (2000). Millenium Development Goals and Beyond 2015. Available at https://www.un.org/millenniumgoals. Accessed on 16 May 2021.

United Nations (2015a). Paris agreement; United Nations framework convention on climate change. Available at The Paris Agreement | UNFCCC. Accessed on 4 January 2021.

United Nations (2015b).The sustainable developmental goals. United Nations general assembly. Available at The 17 Goals | Sustainable Development (un.org). Accessed on 20 September 2020.

United Nations (2015c).Transforming our world:The 2030 Agenda for Sustainable Development. Department of Economic and Social Affairs. Available at https://sdgs.un.org/2030agenda. Accessed on 15 May 2021.

United Nations (2015d). United Nations Sustainable Development Summit 2015. Available at https://www.un.org/sustainabledevelopment/blog/2015/09/summit-charts-new-era-of-sustainable-development-world-leaders-to-gavel-universal-agenda-to-transform-our-world-for-people-and-planet. Accessed on 15 May 2021.

Usher, K., Durkin, J., and Bhullar, N. (2019). Eco-anxiety: How thinking about climate change-related environmental decline is affecting our mental health. *International Journal of Mental Health Nursing* 28 (6): 1233–1234. doi:10.1111/inm.12673.

Vins, H., Bell, J., Saha, S., and Hess, J.J. (2015). The mental health outcomes of drought: A systematic review and causal process diagram. *International Journal of Environmental Research and Public Health* 12 (10): 13251–13275.

Waring, S.C. and Brown, B.J. (2005). The threat of communicable diseases following natural disasters: A public health response. *Disaster Management & Response: DMR: An Official Publication of the Emergency Nurses Association* 3 (2): 41–47. doi:10.1016/j.dmr.2005.02.003.

WHO (2016). International health regulations (2005), 3rd e. World Health Organization. Available at https://apps.who.int/iris/handle/10665/246107. Accessed on 26 September 2020.

WHO (2019). Health Emergency and Disaster Risk Management Framework. Available at https://www.who.int/hac/techguidance/preparedness/health-emergency-and-disaster-risk-management-framework-eng.pdf?ua=1. Accessed on 26 September 2020

WHO (2021) Climate change Available at https://www.who.int/health-topics/climate-change#tab=tab_1. Accessed on 13 March 2021.

WHO Technical Guidance Notes on Sendai Framework Reporting for Ministries of Health (2020). ISBN 978-92-4-000371-2. Available at https://apps.who.int/iris/bitstream/handle/10665/336262/9789240003712-eng.pdf. Accessed 20 February 2021.

World Meteorological Organisation (2014). Risk information: Documenting loss and damage associated with natural hazards and extreme climate events. Bulletin No. 63 (2). World Meteorological Organization (World Meteorological Organisation). Available at https://public.wmo.int/en/resources/bulletin/risk-information-documenting-loss-and-damage-associated-natural-hazards-and. Accessed on 26 September 2020

World Meteorological Organisation (2019). World meteorological congress: Abridged final report of the eighteenth session. 3–14 June 2019, Geneva. World Meteorological Organization (WMO). Available at https://library.wmo.int/doc_num.php?explnum_id=9827. Accessed on 26 September 2020

World Meteorological Organisation (2020) Mandate Available at https://public.wmo.int/en/our-mandate. Accessed on 26 September 2020

7

The Health Costs of Hydrometeorological Extreme Events

Gerardo Sanchez Martinez[1] and Paul Hudson[2]

[1]*The UNEP DTU Partnership, Copenhagen, Denmark*
[2]*Institute for Enbvironmental Sciences and Geography, University of Potsdam, Germany*

7.1　Introduction

Extreme weather events, including floods, storms and heatwaves, can result in large economic impacts (Schmitt et al. 2016), with estimates of global annual impacts range from \$94 billion to over \$130 billion globally (Kousky 2014). Most evaluations and research have focused on understanding the physical impacts (e.g., to assets, capital and property) (Schmitt et al. 2016). However, Hydrometeorological Extreme Events (HEEs) can cause or aggravate a wide range of adverse impacts on health, through complex causal pathways mediated by a great variety of physical and social determinants of health, effect modifiers and influences. For example, extreme heat can cause impacts from mild exhaustion to full heat stroke, aggravate pre-existing chronic conditions and mental health illnesses, increasing emergency hospital admissions as well as premature mortality of most non-accidental causes. Similarly, flooding can cause mortality and injuries directly from drowning, submerged objects, building collapse and moving debris and electrocution; water-, food- and vector-borne disease outbreaks; chemical poisonings; healthcare disruption; and both short-term and long-term mental health disorders (WHO 2013). Moreover, a range of socio-economic factors influence vulnerability and risk from HEEs, which in turn physically depends on daily activities and behaviours, the housing stock and the urban landscape (Bouchama et al. 2007; Bhaskaran et al. 2009; Hajat et al. 2014; Conlon et al. 2011; Martinez et al. 2019; WHO 2008). Together, these factors can generate health impacts, which can be substantial in both immediate and delayed impacts to individuals and society, which are expected to worsen due to climate change (Schmitt et al. 2016). Summarising those figures is beyond

Hydrometeorological Extreme Events and Public Health, First Edition. Edited by
Franziska Matthies-Wiesler and Philippe Quevauviller.
© 2022 John Wiley & Sons Ltd. Published 2022 by John Wiley & Sons Ltd.

the scope of this chapter, but the other chapters elsewhere in this book illustrate them. The complex causal pathways from disasters to health, social and economic impacts are illustrated in Figure 7.1

Therefore, assuming that health effects of HEEs do not imply significant costs or entail severe economic consequences for the communities that have suffered an HEE is undoubtedly reckless. Yet the health costs of HEEs are not usually part of their economic impact assessments or risk management planning. To some extent, this gap reflects scientific uncertainty in the causal chain between HEEs, different health outcomes and resulting economic consequences. A recent study (Horney et al. 2019), analysing the impact of HEEs on Medicare (a health insurance scheme for retirees) costs in US gulf coast states, the researchers found that healthcare expenditures decreased with increased hazard exposure, and that the decrease persisted in the years after the disaster. The authors theorised that this seemingly counterintuitive finding (replicated in other studies) could reflect either a post-disaster limitation in access to healthcare, substitutions in the type of care (i.e., home-based against inpatient) or delay of services. Moreover, studies encompassing the entire US population found that healthcare expenditures increased with disaster severity (Rosenheim et al. 2018). The inconsistency of these results reflects the urgent need for better data and cross-disciplinary analysis in this research area, unfortunately a recurrent theme in the topic of health in Disaster Risk Reduction (DRR).

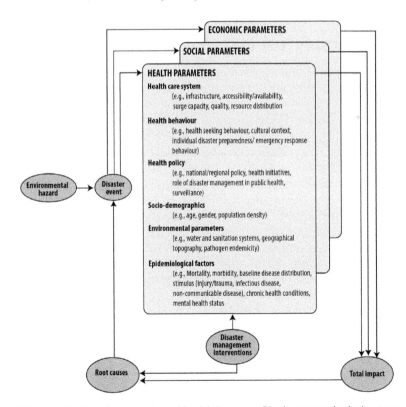

Figure 7.1 Social, economic and health impacts of hydrometeorological extreme events. *Source*: Based on a figure presented in Phalkey and Louis (2016).

Despite the difficulty in their estimation, healthcare costs are arguably the most straightforward type of health-related costs resulting from exposure to HEEs. Estimates of healthcare utilisation, as well as of costs of medications, tests, physician visits or hospital stays are available in most settings, and secondary estimates in publicly available databases exist for areas with patchy data coverage. However, healthcare costs are just one dimension of the direct economic consequences of the health impacts from HEEs. Other indirect but clear costs include loss of income from inability to work, absenteeism from having to care for others, and imperilled livelihoods. Beyond that, death, illness and disability from HEEs affect society and economic systems, reducing output through reduced productivity, depleted capital and reductions in the labour force. In addition, there is the welfare enhancing, though intangible, economic value that individuals and communities place in staying safe and healthy by reducing avoidable risks. All these health economic costs from HEEs can and should routinely be estimated either in part or in full with well-tested current economic tools and methods. This effort can be supported by a small but rapidly growing pool of studies that specifically evaluate the health costs of HEEs. In Section 7.2, we provide an overview of some of the most commonly utilised methods, as well as a discussion on economic considerations regarding the economics of DRR as applied to HEEs and health, and various actionable policy recommendations.

7.2   Estimating the Economic Costs of Hydrometeorological Extreme Events

Estimating and understanding the impacts, including economic ones, of HEEs is important because then we can better design risk management strategies for a more resilient society (Bubeck et al. 2017). Moreover, our understanding of HEE impacts is determined by the data that we collect. There are several commonly used databases for recording the HEE impacts, e.g., those produced by EM-DAT, Munich Re or Swiss Re (Botzen et al. 2019). However, the inclusion criteria for these datasets could be problematic for certain studies (Botzen et al. 2019). Moreover, these databases focus on tangible losses. For instance, health impacts are not recorded as part of the EM-DAT methodology. Similarly, the Post-Event Review Capacity, pioneered by the Zurich Flood Risk Alliance or the Post Disaster Needs Assessment of the World Bank, attempt to forensically study the impacts of a disaster event but do not systematically include health impacts, thereby indicating that the health costs of HEEs may not be systematically recorded in the data we use to underpin our HEE economic assessments. Therefore, there is a strong drive to consider the health impacts of flooding, storms and other HEEs when conducting their risk assessments. Doing so was, for instance, noted in the European Union's Floods Directive (EU 2007). However, no guidance on accounting for these impacts was provided, while monetary flood risk guidance was. Therefore, it is unsurprising that the resulting risk management plans submitted did not actively consider health impacts[1]. A similar finding has also been reported concerning climate change in IPCC reports (Schmitt et al. 2016).

This is problematic because while there are a range of possible decision frameworks for HEEs, cost-benefit analysis (CBA) has become a core decision-making tool (Haasnoot et al. 2013; Mechler 2016; Kunreuther et al. 2013) as it is legally required in some policy-making cases. This is because the CBA rationale is that risk management strategies need to generate a net social benefit in order to be employed, because we only have limited resources and their use must be justified. A CBA seeks to show this by organising, in monetary terms, the costs and benefits (Kunreuther et al. 2013) due to HEE management strategies to proxy the resulting societal welfare change.

The role of CBA in decision-making has, in part, contributed to catastrophe or risk modelling and analysis being a driving force in how we prepare for HEEs (HMT 2011). When designing a risk modelling approach to estimate the costs of a disaster event, it is important to consider what will be included as a cost, as what is not measured cannot be managed (ADB Bank AD 2013). However, the current focus of such models is on the tangible and direct monetary impacts. This is likely because of the disciplinary focus from which catastrophe risk modelling has evolved. This more engineering lead focus tended to measure the tangible impacts, as these impacts were the most intuitive to measure. Extending the range of impacts considered in risk analysis would provide a fuller view of the potential impacts and as such a fuller view of how welfare changes. However, this involves an extension of the range of techniques currently employed.

7.2.1 HEE Risk Assessment and Data

The core element of a risk assessment or analysis is producing a value for risk, i.e., the probability weighted average impact of all possible HEEs, generally for one class of HEE (e.g., fluvial flooding, storm surges, hurricanes or heatwaves). The economic value that comes from avoiding this impact is often the primary source of risk management benefits in CBA. The commonly employed framework of these models is that, broadly speaking, HEE risk (R) is the integral of a function consisting of the hazard event which occurs with a given probability $[H(p)]$, the quantity or value of what can be impacted during an event known as exposure (E), and vulnerability (V) which is how susceptible the exposed are to being harmed during an event. See Kron (2005) for more details. This approach can be fairly accurate if well developed. For instance, Aerts 2014 predicted that Hurricane Sandy inflicted a loss of \$4.2bn, while reported costs were \$4.7bn.

$$R = \int F\big(H(p), E, V\big) dp$$

There has been a great deal of research into the topic concerning the tangible monetary impacts of HEEs (Kron 2005; Kreibich et al. 2014; Thieken et al. 2005; Winsemius et al. 2016; Feyen et al. 2011). In this way, the extent of the Hazard is modelled spatially, Exposure is determined by the value for a given building or land-use in the risk-prone area, and Vulnerability determined by stage-damage curves. Stage-damage curves establish a relationship between HEE intensity and the monetary damage inflicted. Protection standards or infrastructure, such as levees or dams, can be included in this framework by altering the range of probabilities considered as causing damaging events. Together, this produces a value for risk that is also known as the expected annual damage (EAD). Exposure can also be measured as geo-referenced population density or patterns, so that instead the impacted EAP would also be estimated.

While there are continual improvements to the modelled relationships, concerning their accuracy and predictive value, this excludes indirect damage, such as business interruption, which are second-order consequences of an HEE which require different modelling frameworks (Jongman et al. 2014). Moreover, intangible impacts, direct or indirect, are often excluded from risk modelling. This is because they do not have pre-existing monetary values and can be hard to estimate in monetary terms (Alfieri et al. 2018) in the absence of pre-existing markets. Therefore, despite a growing recognition about the large magnitude of intangible impacts (Hudson et al. 2017; Koks et al. 2016; Prettenthaler et al. 2015), they are excluded from the assessment and resulting CBA. Therefore, our current approaches to the CBA of risk management strategies could underestimate the total benefits to society from active risk management.

The health impacts of HEEs can be considered as intangible impacts as they are not directly measurable in monetary terms, though the healthcare costs of treating them is both direct and relatively easily measurable. Additionally, the health impacts from HEEs can themselves be both direct (e.g., injuries suffered during a flood) or indirect (e.g., breathing troubles due to mould growth). Moreover, while it has long been recognised that there is a link between health and various HEEs, e.g., flooding (Hudson et al. 2019a), the complexity of these impacts limits our knowledge on health impacts as compared to monetary costs (Fernandez et al. 2019) leading to a field that while growing remains under researched (Fernandez et al. 2015). However, establishing such a robust data leading understanding across all relevant disciplines is required.

The range of health impacts (both direct and indirect) can be split into two broad camps. The first are physical health impacts typically experienced during or shortly after the event, such as injuries suffered during the flood or storm, infectious diseases due to still water, or those due to water contamination. The second are mental health impacts which may occur around the event or long after its occurrence; these may include severe stress, anxiety and depression, among others (Mambret et al., 2019; Zhong et al., 2018; Saulnier et al. 2017). However, in terms of integrating these impacts into risk assessment would be a slight extension of current practice, it would require a significantly new data baseline.

The specific relation between HEE intensity and different health consequences and their severity are not yet well understood, though it is a rapidly growing field of research (Schmitt et al. 2016). For instance, there are no databases collecting quantitative information on the broader health impacts of flooding (Bubeck et al. 2017), arguably the most prominent HEE to impact humanity (CRED and UNISDR 2015). This is important because without this information it would be impossible to fully integrate these impacts into risk modelling. As an example, Waite et al. (2017) note that there is a positive relationship between the physical flood event and the likelihood of suffering mental health consequences. Therefore, Figure 7.1 presents a simple method for including these health impacts within the common framework for risk assessment. Figure 7.2 (Panel A) indicates that physical losses are often assumed to increase with HEE impacts at a decreasing rate until a maximum level of impact is reached. A similar approach can be applied to health impacts as well.

As an illustration, a given magnitude of a flood event, say 0.5 m inundation depth, is found to increase the incidence of depressive disorders by 1% (epidemiological data), with an average treatment course cost of €2000 in net present value (NPV). NPV is the

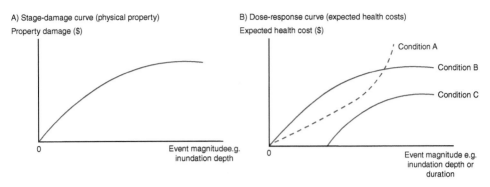

Figure 7.2 A comparison of stage-damage curves. Panel A is the current stage-damage curve for property impacts, while Panel B represents what the health version could look like for a given illnesses, e.g., a mental health outcome assessed through a validated questionnaire, etc. (*Source*: Produced by authors).

total costs of the treatment course over time discounted into a value in today's terms so expenditures at different points in time can be readily compared. Then this event is expected to impact 1000 people, which would imply an aggregate expected cost of 1000 * (1%) * €2000 = €20,000 in additional health impacts for a given HEE. A more extreme flood event, say 1 m inundation depth, is found to increase the incidence rate to 2%, which if impacting 1500 individuals would result in a total NPV of 1500 * (2%) * €2000 = €60,000. This can be added on to the tangible property damage show in Figure 7.2a. Moreover, it operates on a similar assumption to that made for tangible property damages, in that there is a maximum potential loss which is increasingly likely to be reached in the worse event. However, while this is a pragmatic method, it does require several simplifying assumptions that lose a degree of nuance. For instance, for different health conditions, there can be varying event magnitude ranges which are not associated with an incidence probability until a threshold has been reached. These nuances act as variations upon the basic approach, but require additional research on how to understand how generalisable and different these patterns and approaches are. This requires a sufficient robust relationship but in turn would add an additional layer of uncertainty to risk estimates. In order to focus this research, much like with the complex variable selection process involved in the HEE risk management CBAs, it may be productive to focus on the most pressing, likely or important health impacts first as a top research priority.

7.2.2 Methods for Valuing Health Impacts in Monetary Terms

There are a range of techniques that can be used to monetise health impacts in order to both assess their costs and to better integrate these costs into the risk assessments once a sufficiently robust link has been established. Trautmann et al. (2016) notes three categories: Human capital, Economic Growth and Value of a Statistical Life. In addition, there is a growing use of the Life Satisfaction or Subjective Well-being (SWB) approach (Fujiwara and Dolan 2014; OECD 2018). The different methodological approaches indicate different priorities and capture different sets of potential costs or negative impacts. This is relevant as the scoping review of Schmitt et al. (2016) indicates that it depends on the perspective of the analysis on how the costs of health impacts were assessed; for example, when the primary focus of the analysis is on the healthcare system the human capital approach is focused upon. This is because, noted below, it focuses on the direct costs incurred by the health sector while underestimating the costs associated with losses in the quality of life, both of which are relevant aspects of the welfare impact from HEEs which must be measured.

7.2.2.1   The Human Capital Approach The human capital or cost of illness approach involves assessing the direct, or visible, and indirect, or invisible, costs of the illness (COI) (Guh et al. 2008; Schmitt et al. 2016). Direct costs include, for example, medication, physician visits, psychotherapy sessions and hospitalisation (Guh et al. 2008; Jo 2014; Trautmann et al. 2016; WHO 2003). Indirect costs are associated with income losses due to the inability to work or lower levels of productivity (Jo 2014; Trautmann et al. 2016; WHO 2014). This implies an individual level focus for cost assessment. See Larg et al. (2011) for more details.

Concerning direct costs, the human capital approach relies on two types of data: epidemiological and costs (Trautmann et al. 2016). Data can be derived from routine statistics on the costs of treating a given illness such as the average cost of a hospital stay or suitable treatment courses. The particular costs will differ across illnesses and provide an indication of the cost of treating a single person. Epidemiological data relates to the prevalence, the rate of healthcare employment, mortality and length of illness period, in relation to HEEs experiences. This data can be collected based on representative survey data of the target population, through both cross-sectional and longitudinal studies, which link the occurrence and magnitude of HEEs with the relevant health concerns. An aggregate cost estimate is the product of the individual cost data and the epidemiological data.

The advantages of this approach consist of the ease at which it can be integrated into pre-existing cost-benefit analysis. This is because it is based on establishing a deviation from the normal condition to assess the impact. An additional advantage is the relative simplicity of the approach. This is because the relevant information can be drawn from a suitable body of literature or established treatment costs where rules exist (such as the National Health Service in the UK). Moreover, this approach is similar to one of the principles used in the assessment of building damage, in which repair or reconstruction costs are used.

One of the disadvantages of this approach is that healthcare is a highly individualised process and as such the required degree of aggregation causes a loss of nuance. Additionally, it is likely that to establish the actual costs will depend on the extent of the event and for how long treatment was required. This means that the true health costs will only materialise long after the event, which further complicates the analysis as it becomes more difficult to establish which part is due to the HEE or would have occurred due to other circumstances. Moreover, there are criticisms that by focusing on the aggregate costs of treatment, the approach does not connect directly with the value people attach to their lives (see Tarricone 2006, for a discussion). Therefore, Tarricone (2006) argues that this method of cost assessment is best used as a descriptive tool to inform management functions and decision processes. This can be further complicated as new treatment methods can be developed altering cost estimates, which could rapidly invalidate the risk estimates if the approach is not regularly updated. Moreover, it is focused on one illness at a time. This means that it can become quite complicated to include multiple potential in the manner proposed in Figure 7.1, unless a supposed representative bundle of illness is used. However, this could limit the transferability of findings/methods across regions, not only because of regional differences but also potential epistemological uncertainty in the relationship developed.

7.2.2.2 The Economic Growth Approach

The economic growth approach assesses the costs of illness as the total reduction in economic output due to the illnesses (Trautmann et al. 2016). This approach is based on the concepts from economics that economic output is determined (in part) by three factors: productivity, capital stocks, and labour force (Acemoglu 2009). Productivity changes can occur as illness can prevent people from working to their full capacity. Capital stocks can be deleted if ill people must use their savings to finance their healthcare expenditure; lower capital stocks result in fewer investments taking place. Labour falls if ill people are unable to work, and the fewer the workers, the less can be produced. These three avenues will result in less worse overall macroeconomic outcomes. For instance, Chisholm et al. (2016) project the impacts of depression and anxiety by assuming differences in labour force participation and productivity. Therefore, this approach operates at the macroeconomic level.

The economic growth approach loss assessment requires a calibrated economic model of the nation or region that is to be studied, which is combined with actual or simulated HEEs impact data. Botzen et al. (2019) provide a detailed review of how the economic impacts of natural hazards could be assessed. Then this model must be combined with wider models of HEE impacts, epidemiological data, and then the relationship between an HEE's health impact and the consequences on labour force participation or productivity. The difference between the trajectory with and without the HEEs is the cost to society. However, health impacts will form only part of this change.

One advantage of this approach is that it captures a range of costs that are not tied to individuals, but rather systematic indirect economic impacts. These costs are relevant because they indicate how the health status of those impacted can result in negative impacts on others. A second advantage is that economic impacts such as changes in GDP are, in effect, the language of policy-makers. This approach therefore places the relevant impacts in a manner which increases the likelihood of integrating these impacts into the decision-making process and financing adaptation or mitigation strategies. A third advantage is that when the health impacts of HEEs are considered, there is a growing literature investigating the economic impacts of HEEs (Galbusera et al. 2018; Husby et al. 2017; Okuyama and Santos 2014) as random shocks in a range of economic models. Therefore, there is an existing knowledge base which can be extended. However, to the best of our knowledge, these impacts are not currently considered in the approaches developed within HEE research.

The disadvantage of this approach is that it is most suitable for large-scale events which are spread across a nation, or impact major economic hubs, in order to have noticeable impacts on economic activity. This in turn implies that certain HEEs may be more suitable for the economic growth approach than others. For instance, events such as the 2003 European Heat Wave are spread across entire nations. Therefore, it is easy to see how a lower level of national productivity could result in a noticeable reduction in economic activity (Heal and Park 2016). However, floods tend to be much more localised events (Botzen et al. 2019). While it is still true that a flood would destroy a capital, and through illness cause a reduction in labour and productivity, these impacts will be geographically concentrated and other areas would be unaffected. These impacts may not be significant enough at a national level to be noticeable. This is particularly relevant to acknowledge as economic models may operate best at larger scales, while risk models perform better when localised. For this reason, much of the literature exploring these wider economic impacts uses more spatially defined models such as regional input–output models or computable general equilibrium models (Koks et al. 2016). However, this in turn introduces problems as they indicate how to a degree economic activity is transferable between regions. If one region has a lower productive capability because of HEEs but another does not, then a certain amount of economic activity can be transferred to the unaffected region, which is what regional input–output models capture from an HEE shock (Koks and Thissen 2016). The same would be true if these models were extended to include HEE-related health impacts. This reduces the clarity of the health costs, as when the total impacts are aggregated, we see that there is the potential for the negative impacts to be off-set by the positive impacts elsewhere. However, this concern is limited if we only focus on the regions affected, but then the additional insights of this modelling approach is lost. Moreover, the same logic can also be applied to the health industry itself after an event. The cost to an individual because of their ill health after an HEE is a potential gain to a health-care provider and their suppliers.

Therefore, this approach implicitly assumes that gains and losses across actors can be exchanged as in common with CBA (HMT (Her Majesty's Treasury) 2011). Additionally, it can, in principle, generate net positive impacts from HEEs. This is because while someone's property can be destroyed, the cost of them rebuilding can generate a positive economic stimulus. However, this approach increases the ease at which the outcomes can be integrated into our decision-making processes. It focuses on second-order impacts of the HEE, which are relevant impacts. Given the current status of the integration of these approaches, though, this will likely be left to future research, rather than being directly included in the method presented in Figure 7.1.

7.2.2.3   The Value of a Statistical Life   The Value of a Statistical Life (VSL) or Willingness to Pay (WTP) approach assumes that there is a trade-off between the likelihood of suffering an impact and the money spent on reducing this likelihood (Trautmann et al. 2016). While originally developed concerning mortality risk, the general concept is transferable. Therefore, in relation to health cost estimates it can be considered as the total value of money a person is willing to pay to change the incidence rate of a risk or to achieve a specific health outcome (Clark et al. 2014; De Bekker-grob et al. 2012; Mulvaney-Day et al. 2005). Therefore, it is more suitable to consider this as the WTP approach for cost assessment. This is because WTP concepts have often been used to establish values for goods, experiences or services which do not have clear or pre-existing monetary values, especially in relation to environmental issues (Hanley et al. 2003) and public health (Marsh et al. 2012). This approach takes an individual level focus.

The application of this method requires a suitable study of potential WTP values for the range of plausible health impacts. Then the average value for the WTP value can be taken and used in the study. The most common approaches for estimating these values would be via contingent valuation or discrete choice experiments. These methods ask respondents a series of choices regarding their preferences, e.g., $100 per year to reduce the probability of suffering from anxiety disorders by 1% in the wider population, in order to establish how much they are willing to pay for a given change, from which a total value can be inferred. Moreover, if correctly designed, these studies can allow for the true preferences to be uncovered (Mangham et al. 2008).

The advantage of this approach is that it produces actionable values that can be useful in understanding the costs and how much people are willing to pay to address the potential negative impacts by capturing their preferences (Soeteman et al. 2017). Hence, this approach is eminently suitable for policy-makers or risk managers, as they could see a rationale for the use of tax money or requesting contributions towards a fund. Therefore, it is readily useable in risk management strategies (Botzen et al. 2009; Brouwer and Van Ek 2004; Devkota et al. 2018). Moreover, in comparison to the human capital approach, it has been argued that the WTP measures are the more theoretically correct measure (Guh et al. 2008). This is because the WTP includes a wider view of the perceived disutility of illness and as such welfare changes, which is the cost that socially optimal decision-making should be based upon.

There are also several disadvantages of this approach. The first is that it can be difficult to establish a suitable WTP value for a region or population segment without a suitably resourced study beforehand. To avoid this, many projects use values-transfer where a WTP value from a different study area is used instead. However, these values are highly context dependent. For instance, in a study related to environmental preferences, errors of up to approximately 75% (Brander et al. 2006) could result from values-transfers, resulting in highly inaccurate cost estimates. A disadvantage related to context dependence is timing, as the stability of preferences over time is unclear (Brouwer et al. 2017; Neher et al. 2017). However, this might not be the case, as it is known that people's preferences regarding HEEs can change in the wake of experiencing HEEs. Additionally, the estimated preferences could also reflect hypothetical bias (Ryan et al. 2017). This is because we are asking respondents to compare and evaluate choices of something they may not have tangible experience with, so this means that they do not reflect suitably on their own preferences to suitability value them. Moreover, this problem is also related to strategic (Meginnis et al. 2018) or protest (Rakotonarivo et al. 2016) choices, whereby people do not reveal their preferences but act to achieve a specific aim rather than reflecting their preferences.

While WTP estimates tend to be focused on specific illnesses, to increase the tangibility of choices, they can also feasibly look at changes in health status which could be considered as a holistic understanding of potential health changes.

7.2.2.4 The Subjective Well-Being or Life Satisfaction Approach Additionally, there is the subjective well-being (SWB) or life satisfaction approach (Brown 2015;

Howley 2017; Huang et al. 2018; Powdthavee and Van Den Bergh 2011). This approach argues that asking individuals to state their SWB provides a direct proxy of welfare (Frey et al. 2009) and the full range of impacts suffered (Fujiwara and Dolan 2014). The relationship between health and well-being and the separate relationship with income and well-being are combined to achieve the monetary value of the impact. This is often the amount of money required to off-set changes in well-being. There is a growing literature on the SWB impacts of HEEs (Hudson et al. 2017, 2019a; Fernandez et al. 2019; Luechinger and Raschky 2009; Sekulova and van den Bergh 2016; Von Möllendorff and Hirschfeld 2016), which has focused so far on establishing a general reduction in welfare. This is in line with the approach suggested by Fujiwara and Dolan (2014) and Lamond et al. (2015) and is slightly different from the medical literature which has also studied specific health conditions (e.g., Powdthavee and Van Den Bergh 2011).

The SWB approach has several advantages. The first is that SWB data allow us to determine what matters in people's lives when they are not thinking about how much those things matter. Regression analysis determines the importance of different health states and the other determinants of SWB (Dolan et al. 2012). This approach establishes indirect relationships, removing an incentive to directly understate their status by limiting the degree of strategic and hypothetical bias (OECD 2018). When compared to WTP approaches it is an additional benefit of SWB approaches that strict assumptions on the rationality of preferences is not required as they directly measure welfare (OECD 2018). Additionally, the SWB approach is based on actual rather than hypothetical statements, thereby concentrating on lived experiences in addition to predicted responses regarding their impacts. Fujiwara and Dolan (2014) argue that SWB data may provide a better representation of how people are affected by health conditions than stated preference methods that are used as part of measuring Quality Adjusted Life Years (QALYs). However, the SWB approach is better suited to valuing non-marginal changes, e.g., negative mental health following a flood rather than the risk of negative mental health after a flood.

There are also several limitations to the SWB approach. One limitation is that the link between income and SWB is important as otherwise losses may be incorrectly estimated (Ambrey et al. 2014; OECD 2018). The way SWB is measured presents another limitation. While life satisfaction is the most commonly used aspect of SWB, eudaimonic or momentary well-being could also be used. These different aspects of SWB may have different determinants and display different levels of sensitivity to impacts (OECD 2018). These problems can be further compounded by using too narrow Likert scales and only asking one single overall question given the complexity of the problem the respondent would face (OECD 2018). Finally, a further issue is the transferability of findings from one region to another, much like WTP, due to its inherent subjective nature.

SWB estimates can be readily focused on changes of general health status, as SWB with health is an often-employed SWB question, which could capture the well-being effects of multiple health conditions. However, establishing a link with the probability of suffering such a health impact might be more difficult than the other measures discussed as it is less tangibly focused.

7.2.3 Projections of Changes in Health and Well-Being Costs of Hydrometeorological Extreme Events Under Different Climate Change Scenarios

Climate change is widely projected to increase the frequency and intensity of most HEEs globally. However, projecting whether and how that increase may translate into health impacts constitutes a formidable conceptual and practical challenge, for several reasons. Crucially, there is not a simple or universal definition of an HEE or of most of the specific types of events therein. In addition, measuring the health impacts of HEEs is complex in practice, varying according to the particular HEE and the type of health outcome (e.g., immediate, like injuries or drowning, or delayed, like long-term mental illnesses). These challenges apply even in settings with good data availability and functional and efficient information systems, a situation rarely found, particularly in developing countries. Given this baseline situation, projections of health impacts of HEEs under climate change are fraught with uncertainty, compounded by the inherent complexity of alternative social, economic, environmental and climate scenario building, as was the case with the risk modelling of HEE health impacts.

Based on these already uncertain projections, the calculation of prospective health economic costs of HEEs adds an additional layer of unpredictability regarding a wide range of factors, from macroeconomic trends such as growth and employment to societal and individual preferences and behavioural aspects. Unsurprisingly, again, this results in an under-representation of health costs in the projections of economic costs of HEEs under climate change.

A relative exception is heatwaves, where a comprehensive pool of studies and evidence on the relationship between temperature and health combined with the higher reliability of downscaled long-term temperature projections (as compared to humidity and precipitations, Kirtman et al. 2013) provides for a relatively solid basis on which to project certain types of health costs under climate change. For instance, a study of 23 countries projected that warmer regions would experience a sharp surge in heat-related mortality from around 3% in Central America to around 13% in South-east Asia by the end of the century (Gasparrini et al. 2017). Based on similar types of modelling, the welfare and GDP impact of heat-related health outcomes under climate change has been calculated for the EU (Ciscar et al. 2014). Besides global and regional assessments, downscaled climate models have allowed projecting heat-related mortality for hundreds of specific urban settings, though mostly in rich countries (Campbell et al. 2018).

Additionally, the scoping review by Schmitt et al. (2016) further illustrates that the health-related economic burden of extreme events is likely to increase steeply under projected increases in mean temperatures globally, a projection that entails increasingly far-reaching consequences as evidence emerges of: a) the longer latency period of some HEE health outcomes; b) the heavily socially mediated nature of vulnerability to HEEs; and c) the wide income-related disparities in vulnerability.

7.3   Reducing or Off-Setting the Health Costs of Hydrometeorological Extreme Events

The health costs of HEEs provide a strong incentive for their active management and DRR in order to limit the burden HEEs place upon society. This incentive naturally fits within the calls made as part of the Sendai Framework for Disaster Risk Reduction (UNDRR 2015) and the Sustainable Development Goals (UN 2015), for instance, in reducing the impacts of HEEs (see Chapters 5 and 6).

Acting upon these calls has resulted in a growing focus on understanding how people will respond to and recover from their HEE experiences, and how this can be used as part of HEE risk assessments (Aerts et al. 2018). This is because of how patterns of property-level adaptation measure employment change with event experiences. Moreover, in employing these measures, individual well-being can increase (Hudson et al. 2017) or lower the likelihood of negative mental health disorders if the impact of the HEE is reduced (Paranjothy et al. 2011) in addition to physical damage reduction. There are many DRR measures outside the health sector that those affected can use to lower the impacts of an HEE in a cost-effective manner. For example, in the case of flooding, there are a range of measures to keep water out of a building, known as dry flood-proofing, or to limit impacts once water has entered a building, known as wet flood-proofing. (Hudson et al. 2014; Joseph et al. 2015; Kreibich et al. 2015; Lamond et al. 2018; Poussin et al. 2015). In lowering the impact of an event, via DRR, we reduce the effective incidence rate of health consequences. This was found to be the case in Greater Houston after Hurricane Harvey (Grineski et al. 2019). Additionally, a study in the UK found that, aside from the known effects of flood-related displacement on mental health outcomes, both the timing of the displacement and the amount of warning received in advance could modify the severity of the outcomes in the long term (Munro et al. 2017). The integration of health impacts into risk assessment will increase the effective benefits of DRR. This in turn will result in a wider range of DRR measures being found to be cost-effective. Moreover, people at lower risk of flooding may find DRR becomes effective when a wider range of potential benefits are considered. However, the health protection potential from DRR should not be automatically assumed, but scrutinised based on the best available evidence, due to individual nuances of various health conditions.

Despite this, from the standpoint of individual actors (e.g., a healthcare facility) the benefits of investing to reduce the economic risks associated to HEEs may sometimes be unclear. To begin with, like other risk management strategies, there is not an immediate economic saving or financial incentive to reduce the risks associated to HEEs: it does not affect the business of treating patients, which is what ultimately drives most decisions in the sector. In addition, different risk profiles according to location and other factors may make HEE-proofing a cost-beneficial decision in one setting and not another. At most aggregated levels, however, reducing or off-setting the costs of HEEs often makes good financial sense. From the collective standpoint of health systems and healthcare providers, adequate emergency preparedness and management of HEEs reduces risks and saves costs. This is, for instance, illustrated by the steep disaster-related costs to hospitals (including capital, operation, emergency work, uncompensated care and other costs) due to Hurricane Harvey in

Texas, which prompted the relevant hospital associations to strongly advocate for DRR investment. The case for investing to reduce health-related economic costs of HEEs is even clearer from a societal perspective, particularly when all HEE-related costs (including non-market and indirect ones) are taken into account. In some cases, researchers have been able to calculate current and projected benefit-to-cost ratios (BCRs) of preventive measures against heatwaves; BCRs were estimated currently at 11 (e.g., Euros saved per Euro invested) for London, 308 for Prague and 913 for Madrid, increasing much further in the near future under all climate scenarios (Hunt et al. 2017).

At the general societal level, there are wider strategies and mechanisms that can be strategically employed to lower the burden of HEE health impacts, as well as a relatively good agreement on the types of activities needed to protect health from HEEs in a changing climate. Predictably, harmful exposures to HEEs as well as climate-sensitive diseases and impacts from extreme weather events and disasters are more frequent amongst the world's poorest populations (Byers et al. 2018a; WHO 2018; WHO and World Meteorological Organisation 2012). Therefore, to a large extent, reducing the risk of health impacts from and increasing health systems resilience to HEEs globally will continue. It will depend largely on basic development activities (UNEP 2019), in particularly those that address key basic social and environmental determinants of health, such as adequate access to safe water and sanitation. In addition, standard DRR activities such as general-level emergency mitigation, preparedness, response and recovery, or specific systems such as early warning systems, are fundamental to reduce or minimise health impacts from HEEs. In order for these activities to maximise their health protective potential, however, health systems should be involved in their planning, development and implementation from inception (see Chapter 6). The integration of the health sector in risk management is an under-appreciated aspect of the common call for multi-sector partnerships for DRR. In addition to general DRR activities, health authorities are increasingly engaging in HEE-specific health prevention plans. A key example of these activities are Heat-Health Action Plans (HHAPs), which have become relatively widespread in developed countries and are increasingly being implemented in developing country settings (Martinez et al. 2019). Other examples include the use of mobile technologies to ensure access to healthcare of vulnerable populations after HEEs, and specific interventions to ensure the possibility of post-event self-care of chronically ill patients, etc. However, the reviews of Schmitt et al. (2016) and Bouzid et al. (2013) find few studies that evaluate the cost-effectiveness of health-related DRR strategies outside of heatwaves. This is a prime source of future research in developing optimal adaptation strategies moving forward.

7.3.1 Increasing the Resilience of Health Systems to Hydrometeorological Extreme Events

In a context of increasingly frequent and severe HEEs (see Chapter 2), health systems must engage in efforts to increase their climate resilience. Recent events have highlighted the vulnerability of healthcare facilities to extreme weather in different settings, and the consequences for the community when these vital service providers are affected. Climate-related risks, and specifically extreme weather events, are expected in general to worsen under most likely climate scenarios, and healthcare providers and administrators both in developed and developing countries are acknowledging the need for healthcare facilities to build the capacity to cope with and adapt to current and projected impacts (Bowen and Ebi 2017). Equipment, know-how and organisational capacity all play a key role in building this resilience, thus making it a prime case study of health-protecting adaptation technology. In its operational framework for climate resilient health systems, WHO (2015a) identifies up to ten key areas of work, from leadership and governance to financing.

One crucial area of work is the climate-proofing of health facilities. In this regard, a great number of initiatives have been taking place at the subnational and national levels aiming at building healthcare facility resilience to extreme weather and climate change. These have happened either in connection with or aligned with multilateral agreements. Health is a central component of the post-2015 framework for DRR; at the Sendai conference, WHO released a Comprehensive Safe Hospital Framework to guide the development and implementation of Safe Hospital programmes at the national, subnational and facility levels (WHO 2015b). In terms of specific activities for hospital safety during emergencies, the Hospital Safety Index Guide (WHO 2015c) outlines some priorities:

- enable hospitals to continue to function and provide adequate care during and following emergencies;
- protect health workers, patients and families;
- protect the physical integrity of healthcare facilities, equipment and critical systems; and
- make hospitals safe and resilient to future risks, including climate change.

The checklist of elements to take into account when evaluating the resilience of healthcare facilities is very comprehensive, including hardware, software and 'orgware' (i.e., the capacity of institutional actors to deliver on goals, in this case related to resilience). It is typically organised around four key dimensions: hazard evaluation; structural integrity; non-structural integrity; and emergency management. Ultimately, strategies for hospital and healthcare facility resilience should be prioritised according to the local circumstances and the best available evidence. For example, healthcare facilities should consider in advance the objective criteria for their decisions of sheltering in place vs. evacuation in case of HEE based on empirical evidence and their own circumstances (Downey et al. 2013; King et al. 2016). Additionally, differing levels of social vulnerability can result in different health impacts, both in the short term and long term (Schmitt et al. 2016). Therefore, different areas will have to undertake different actions to improve their resilience, which account for both their HEE risk profile and their socio-economic circumstances.

7.3.2   The Role of Insurance and Other Tools

Insurance can play a fundamental role in minimising the financial risk associated with HEEs, from the individual level to the societal level. At the individual level, insurance has been proved to be a protective factor against flood-related severe stress and psychological morbidity for instance (Mulchandani et al. 2019).

Insurance fulfils several roles to achieve this reduction in health-related HEE risk. The first role is that affordable insurance can facilitate access to healthcare by reducing the need for potentially large and catastrophic out-of-pocket payments (Mas-Colell 1995; Odeyemi and Nixon 2013), which in turn limits long-term health consequences. The second role is from the observation that the most common role of insurance as a protective action against HEEs is to help the policyholder absorb the financial impact of replacing and repairing the physical property damage caused by the event; for example, replacing carpets, cleaning costs from contaminated water or addressing mould growths. It is commonly argued that well-functioning and designed HEEs insurance markets can provide incentives for additional physical risk reduction behaviour (Hudson et al. 2016, 2019a; Surminski et al. 2015; Surminski and Thieken 2017). An example is the employment of mobile flood barriers before a flood to reduce the likelihood that water enters a building. This additional action should mean that the impacts of a given HEE event are smaller, lowering the likelihood of negative health consequences. Moreover, as noted above, the protection of health infrastructure and maintaining continual activity is paramount. For this, Unterberger et al. (2019) argue that encouraging governments to more formally insure critical infrastructure and to actively link the charged insurance premiums to health infrastructure risk management would build resilience in the health sector.

However, it must be noted that the role of insurance is limited by the ability to access insurance. In countries with access to universal healthcare, this concern is reduced. Where this is not the case, parametric insurance could support access to insurance. Traditional insurance approaches are based on indemnity principles which mean that the insurance company will reimburse the exact loss suffered. However, indemnity approaches can be expensive as, for example, a system of loss adjusters is required, especially in countries without a high level of insurance coverage. Parametric insurance greatly simplifies the insurance process because such insurance provides a pre-set payment if a certain triggering event occurs. For instance, $1000 can be distributed to policyholders if it has not rained for 3 months. This approach is much simpler and quicker, reducing the transaction costs and potentially premiums. However, to the best of our knowledge, parametric insurance has not been directly applied to healthcare markets, while it has a tradition of being employed against HEEs. It is likely the result of the problems of setting a trigger event with health as compared to HEEs. The presence of an HEE can be monitored externally, e.g., through radar data, which may not be the case for more individual healthcare issues. There is also the problem of basis risk, which is where an event inflicts damage, while just being under the trigger and as such no potential pay-outs or an event occurs triggering a pay-out but without major impacts. In essence there is the problem for the insured that they could fall ill but not sufficiently ill to trigger insurance payments. While the problem of basis risk also applies to natural hazard insurance, it has not inhibited potential products being offered. Therefore, there is scope to extend current parametric insurance policies to cover health expenses. However, additional research and engagement with stakeholders will be required to understand how best to achieve this extension.

7.3.3 Innovative Funding Sources

DRR to reduce the health impact of HEEs requires additional expenditure compared to the status quo expenditure patterns of governments and individuals around the world. Individuals can respond to incentives provided to them by a range of higher-level stakeholders, who in turn require resources to provide these incentives. These additional resources can be from re-allocating current spending, increased taxation or borrowing. However, there is also a strong case for directing additional international funding to proactive DRR. Currently, a significant proportion (72%) of Official Development Assistance (ODA) is being spent on emergency response, whereas only a very small proportion (around 4%) is directed towards prevention (Peters 2018). Similarly, the share of international climate funding dedicated to health adaptation is virtually negligible. Between 2003 and 2017, less than 1% of international finance for climate change adaptation was allocated to health (Watts 2018). The presence of health adaptation projects in the relevant multilateral funds (e.g., the Special Climate Change Fund, the Least Developed Countries Fund, and the Green Climate Fund) is either scarce or non-existent. Yet there is a strong case for dedicating climate adaptation funds to, for example, the climate-proofing of health facilities. It is a prime example of health adaptation, since the continuity of care during HEEs strongly increases the resilience of the surrounding community. Given the large human and economic losses associated to hospital failure during emergencies, climate-proofing is likely to be highly cost-beneficial. Beyond the health sector, health adaptation technology is likely to provide large co-benefits to other sectors (e.g., by reducing sick days) (Roschnik et al. 2017). These activities cut across and integrate ongoing multilateral processes, such as the Sendai Framework, the United Nations Framework Convention on Climate Change (UNFCCC) (United Nations 1992) and SDGs (UN 2015).

Additionally, not only can lessons be learnt from current patterns of ODA but also from the various institutions aimed at enabling access to international financing. For instance, the African Risk Capacity (ARC), which supports parametric re-insurance arrangements to allow nations access to international financing for disaster losses which they may not have been able to otherwise support. The ARC in particular helps to support access to financing if the country displays that it has made suitable efforts to actively conduct risk management. Similar structures may be useful for the health systems by providing re-insurance facilities which support local health systems to add additional capacity, in times of need, while promoting additional discipline in the risk management of HEEs-related health impacts. However, the trigger of such a policy would be most likely met after an event has occurred, which means that the money arrives after the effects of an event are felt. Rather, what could be more successful is the employment of forecast-based financing for risk management. Forecast-based financing is based on the idea that funds can be mobilised in accordance with the likelihood of an HEE, so that short-term strategies such as cash transfers (Guimarães Nobre et al. 2019) or evacuation (Coughlan De Perez et al. 2015) can be put in place before a HEE occurs in order to limit its potential impacts. Forecast-based financing can allocate resources to healthcare facilities before an event so that health impacts can be managed as they appear.

7.4   Conclusions and Recommendations

The potential health impacts of HEEs are large in economic and human terms. These costs to human welfare and the resulting economic benefit from reducing their incidence rates needs to be better accounted for in decision-making. This will require actively including stakeholders from the health sector as part of long-term HEE risk management, echoing the common call to establish multi-sector partnerships for DRR while expanding upon the commonly suggested set of insurers, governments and households. The successful integration of the health sector into DRR discussions and the expertise that they bring will aid in the integration of health impacts into risk modelling. Once a robust understanding of the important health impacts has been achieved and converted into useful monetary values, a strong business case for their cost-effective management can be developed. Extending the range of actively considered stakeholders will bring in the new sources of knowledge, experience and data required in establishing this new baseline. However, care will have to be taken in establishing successful collaboration across stakeholders who at first glance may not be immediately natural partners. Even though there are several areas in which the various stakeholders are conducting parallel research that can be connected to true advance on this topic, differences in research language and habits can make collaboration difficult.

With this need in mind, we offer the following suggestions for joint activities for the relevant stakeholders to undertake moving forwards, bridging disciplines and sectors in order to create a community of users of secure, safe and resilient societies:

- More focused inter/trans-disciplinary work is needed to fully integrate health costs into HEE risk management. Cross-cutting measures and research agendas are needed in order to establish a robust data baseline for HEE risk assessments. For instance, the research currently conducted in the medical literature on linking HEE magnitudes to changes in mental health already naturally crosses over with the stage-damage curves employed in risk research. Acting upon this overlap is a natural first step.
- The health impacts are potentially long-lasting and slow onset, as compared to property damage, therefore there needs to be a greater focus on longitudinal data collection (Hudson et al. 2019a) rather than cross-sectional surveys of those impacted by HEEs. The medical community has more experience in this field than flood risk managers. Hence, there is scope for improved collaboration across disciplines to render long-term investments in longitudinal datasets more advantageous.
- Including public health elements more systematically in risk management, so that we can bring in knowledge that can minimise HEE-related health impacts, echoing the common call to establish multi-sector partnerships for DRR. One suitable activity for all these stakeholders to jointly act upon is developing a harmonised impact recording methodological that includes the most relevant human and physical impacts.
- Expanding the scope and coverage of studies conducting economic evaluation of health impacts from extreme weather events, which hitherto have mainly focused on heatwaves and has been mostly focused on the richer areas of Asia and the USA (Schmitt et al. 2016). Better coverage of HEEs and developing countries is urgently needed to inform adequate multilateral DRR and climate change adaptation policy and assistance.
- Society must be suitably prepared for HEEs before they occur. This can be through various prevention or impact mitigation strategies or to have recovery strategies in place to help speed up recovery. Therefore, flexible forecast-based financing can be

useful for allocating funds for addressing the expected public health needs after a disaster event before it occurs. In order to do so successfully, silos between sectors must be broken down. as developing suitable financing strategies would require risk management actors to determine the possible size and occurrence of HEEs and health actors to determine what would be suitable resources for managing the impacts of those events.

Note

1. The submitted responses can be found at: https://ec.europa.eu/environment/water/water-frame work/impl_reports.htm.

References

Acemoglu, D. (2009). *Introduction to Modern Economic Growth*. Princeton, NJ: Princeton Univeristy Press. https://econpapers.repec.org/RePEc:cla:levrem:122247000000001721.

Aerts, J.C.J.H. (2014). Climate adaptation. Evaluating flood resilience strategies for coastal megacities. *Science* 344 (6183): 473–475.

Aerts, J.C.J.H., Botzen, W., Clarke, K.C. et al. (2018). Integrating human behaviour dynamics into flood disaster risk assessment. *Nature Climate Change* 8: 193–199. doi:10.1038/s41558-018-0085-1.

Alfieri, L., Dottori, F., Betts, R. et al. (2018). Multi-model projections of river flood risk in Europe under global warming. *Climate* 6 (1): 6. http://www.mdpi.com/2225-1154/6/1/6.

Ambrey, C.L. and Fleming, C.M. (2014). The causal effect of income on life satisfaction and the implications for valuing non-market goods. *Economics Letters* 123 (2): 131–134. https://doi.org/10.1016/j.econlet.2014.01.031.

ADB, Bank AD (2013). *Cost-Benefit Analysis for Development: A Practical Guide* (ed. B. Ad). Mandaluyong: Asian Development Bank.

Bhaskaran, K., Hajat, S., Haines, A. et al. (2009). Effects of ambient temperature on the incidence of myocardial infarction. *Heart* 95 (21): 1760–1769. doi:10.1136/hrt.2009.175000.

Botzen, W.J.W., Aerts, J.C.J.H., and Van Den Bergh, J.C.J.M. (2009). Willingness of homeowners to mitigate climate risk through insurance. *Ecological Economics: The Journal of the International Society for Ecological Economics* 68 (8–9): 2265–2277.

Botzen, W.J.W., Deschenes, O., and Sanders, M. (2019). The economic impacts of natural disasters: A review of macroeconomic computational models and empirical studies. *Review of Environmental Economics and Policy 13 (2): 167–$88.*

Bouchama, A., Dehbi, M., Mohamed, G. et al. (2007). Prognostic factors in heatwave-related deaths: A meta-analysis. *Archives of Internal Medicine* 167 (20): 2170–2176. doi:10.1001/archinte.167.20.ira70009.

Bouzid, M., Hooper, L., and Hunter, P.R. (2013). The effectivness of public health interventions to reduce the health impact of climate change: A systematic review of systematic review. *PLoS One* 8: 1–15.

Bowen, K. and Ebi, K. (2017). Health risks of climate change in the World Health Organization South-East Asia Region. *WHO South-East Asia Journal of Public Health* 6: 3–8.

Brander, L.M., Florax, R.J.G.M., and Vermaat, J.E. (2006). The empirics of wetland valuation: A comprehensive summary and a meta-analysis of the literature. *Environmental and Resource Economics* 33 (2): 223–250. doi:10.1007/s10640-005-3104-4.

Brouwer, R., Sheremet O., and Logar, I. (2017). Choice consistency and preference stability in test-retests of discrete choice experiment and open-ended willingness to pay elicitation formats. *Environmental and Resource Economics* 68 (3): 729–751. doi:10.1007/s10640-016-0045-z.

Brouwer, R. and Van Ek, R. (2004). Integrated ecological, economic and social impact assessment of alternative flood control policies in the Netherlands. *Ecological Economics* 50: 1–21.

Brown, T.T. (2015). The subjective well-being method of valuation: An application to general health status. *Health Services Research* 50 (6): 1996–2018. doi:10.1111/1475-6773.12294.

Bubeck, P., Otto, A., and Weichselgartner, J. (2017). Societal impacts of flood hazards. In: *Oxford Research Encyclopedia on Natural Hazards Science*.https://oxfordre.com/naturalhazardscience/view/10.1093/acrefore/9780199389407.001.0001/acrefore-9780199389407-e-281. Accessed 6 September 2021.

Byers, E., Gidden, M., and Lecl, D. (2018a). Global exposure and vulnerability to multi-sector development and climate change hotspots. *Environmental Research Letters* 13: 055012. doi:10.1088/1748-9326/aabf45.

Campbell, S., Remenyi, T.A., White, C.J. et al. (2018). Heatwave and health impact research: A global review. *Healing Place* 53: 210–218. doi:10.1016/j.healthplace.2018.08.017.

Chisholm, D., Sweeny, K., Sheehan, P. et al. (2016). Scaling-up treatment of depression and anxiety: A global return on investment analysis. *The Lancet Psychiatry* 3 (5): 415–424. https://doi.org/10.1016/S2215-0366(16)30024-4.

Ciscar, J., Feyen, L., Soria, A. et al. (2014). *Climate Impacts in Europe. The JRC PESETA II Project*. European Commission Joint Research Centre Institute for Prospective Technological Studies.

Clark, M.D., Determann, D., Petrou, S. et al. (2014). Discrete choice experiments in health economics: A review of the literature. *Pharmacoeconomics* 32 (9): 883–902. doi:10.1007/s40273-014-0170-x.

Conlon, K.C., Rajkovich, N.B., White-Newsome, J.L. et al. (2011). Preventing cold-related morbidity and mortality in a changing climate. *Maturitas* 69 (1873–4111 (Electronic)): 197–202.

Coughlan De Perez, E., Van Den Hurk, B., Van Aalst, M.K. et al. (2015). Forecast-based financing: An approach for catalyzing humanitarian action based on extreme weather and climate forecasts. *Natural Hazards and Earth System Sciences* 15 (4): 895–904. doi:10.5194/nhess-15-895-2015.

CRED-UNISDR. (2015). The human cost of weather-related disasters, 1995–2015, CRED and UNISDR, https://www.unisdr.org/files/46796_cop21weatherdisastersreport2015.pdf

CRED and UNISDR (Centre for Research on the Epidemiology of Disasters and United Nations Office for Disaster Risk Reduction) (2018) *Economic Losses, Poverty &Disasters: 1998–2017*. 465. https://www.preventionweb.net/files/61119_credeconomiclosses.pdf

De Bekker-grob, E.W., Ryan, M., and Gerard, K. (2012). Discrete choice experiments in health economics: A review of the literature. *Health Economics* 21 (2): 145–172. doi:10.1002/hec.1697.

Devkota, R.P. and Maraseni, T. (2018). Flood risk management under climate change: A hydro-economic perspective. *Water Supply* 18 (5): 1832–1840. doi:10.2166/ws.2018.003.

Dolan, P. and Metcalfe, R. (2012). Valuing health: A brief report on subjective well-being versus preferences. *Medical Decision Making* 32 (4): 578–582. doi:10.1177/0272989X11435173.

Downey, E.L., Andress, K., and Schultz, C.H. (2013). Initial management of hospital evacuations caused by Hurricane Rita: A systematic investigation. *Prehospital and Disaster Medicine* 28 (3): 257–263. doi:10.1017/S1049023X13000150.

EU (2007) *Directive 2007/60/EC of the European Parliament and of the Council.*

Fernandez, A., Black, J., Jones, M. et al. (2015). Flooding and mental health, a systematic mapping review. *PLoS One* 10 (4): e0119929. doi:10.1371/journal.pone.0119929.

Fernandez, C.J., Stoeckl, N., and Welters, R. (2019). The cost of doing nothing in the face of climate change: A case study, using the life satisfaction approach to value the tangible and intangible costs of flooding in the Philippines. *Climate and Development* 1–14. doi:10.1080/17 565529.2019.1579697.

Feyen, L., Dankers, R., Bódis, K. et al. (2011). Fluvial flood risk in Europe in present and future climates. *Climatic Change* 112 (1): 47–62. doi:10.1007/s10584-011-0339-7.

Frey, B.S., Luechinger, S., and Stutzer, A. (2009). The life satisfaction approach to valuing public goods: The case of terrorism. *Public Choice* 138: 317–345.

Fujiwara, D. and Dolan, P. (2014) *Valuing Mental Health: How a Subjective Wellbeing Appraoch Can Show Just How Much It Matters*. Online only: UK Council for Psychotherapy. https://www.psychotherapy.org.uk/wp-content/uploads/2018/08/UKCP_DocumentsReports ValuingMentalHealth_web.pdf

Galbusera, L. and Giannopoulos, G. (2018). On input–output economic models in disaster impact assessment. *International Journal of Disaster Risk Reduction* 30: 186–198. https://doi.org/10.1016/j.ijdrr.2018.04.030.

Gasparrini, A., Guo, Y., Sera, F. et al. (2017). Projections of temperature-related excess mortality under climate change scenarios. *Lancet Planetary Health* 1 (9): e360–e367. https://doi.org/10.1016/S2542-5196(17)30156-0.

Grineski, S.E., Flores, A.B., Collins, T.W. et al. (2019). The impact of Hurricane Harvey on Greater Houston households: Comparing pre-event preparedness with post-event health effects, event exposures, and recovery. *Disasters* 44: 408–432.doi:10.1111/disa.12368.

Guh, S., Xingbao, C., Poulos, C. et al. (2008). Comparison of cost-of-illness with willingness-to-pay estimates to avoid shigellosis: Evidence from China. *Health Policy and Planning* 23 (2): 125–136. doi:10.1093/heapol/czm047.

Guimarães Nobre, G., Davenport, F., Bischiniotis, K. et al. (2019). Financing agricultural drought risk through ex-ante cash transfers. *The Science of the Total Environment* 653: 523–535. https://doi.org/10.1016/j.scitotenv.2018.10.406.

Haasnoot, M., Kwakkel, J.H., Walker, W.E. et al. (2013). Dynamic adaptive policy pathways: A method for crafting robust decisions for a deeply uncertain world. *Global Environmental Change* 23: 485–498.

Hajat, S., Vardoulakis, S., Heaviside, C. et al. (2014). Climate change effects on human health: Projections of temperature-related mortality for the United Kingdom during the 2020s, 2050s and 2080s. *Journal of Epidemiology and Community Health* 68. doi:10.1136/jech-2013-202449.

Hanley, N., Ryan, M., and Wright, R. (2003). Estimating the monetary value of health care: Lessons from environmental economics. *Health Economics* 12 (1): 3–16. doi:10.1002/hec.763.

Heal, G. and Park, J. (2016). Reflections –Temperature stress and the direct impact of climate change: A review of an emerging literature. *Review of Environmental Economics and Policy* 10 (2): 347–362. doi:10.1093/reep/rew007.

HMT (Her Majesty's Treasury) (2011). *The Green Book. Appraisal and Evaluation in Central Government*. London: Her Majesty's Treasury.

Horney, J., Rosenheim, N., Zhao, H. et al. (2019). The impact of natural disasters on medicare costs in U.S. gulf coast states. *Medicine (Baltimore)* 98 (19):e15589.doi:10.1097/MD.0000000000015589.

Howley, P. (2017). Less money or better health? Evaluating individual's willingness to make trade-offs using life satisfaction data. *Journal of Economic Behavior & Organization* 135: 53–65. https://doi.org/10.1016/j.jebo.2017.01.010.

Huang, L., Frijters, P., Dalziel, K. et al. (2018). Life satisfaction, QALYs, and the monetary value of health. *Social Science & Medicine* 211:131–136. https://doi.org/10.1016/j.socscimed.2018.06.009.

Hudson, P., Botzen, W.J.W., Feyen, L. et al. (2016). Incentivising flood risk adaptation through risk based insurance premiums:Trade-offs between affordability and risk reduction. *Ecological Economics : The Journal of the International Society for Ecological Economics* 125: 1–13.

Hudson, P., Botzen, W.J.W., Kreibich, H. et al. (2014). Evaluating the effectiveness of flood damage risk reductions by the application of Propensity Score Matching. *Natural Hazards and Earth System Sciences* 14: 1731–1747.

Hudson, P., Botzen, W.J.W., Poussin, J.K. et al. (2017). The impacts of flooding and flood preparedness on happiness: A monetisation of the tangible and intangible subjective well-being impacts. *Journal of Happiness Studies*. doi:10.1007/s1. doi:10.1007/s10902-017-9916-4.

Hudson, P., De Ruig, L., De Ruiter, M., et al. (2019a). Best practices of extreme weather insurance in Europe and directions for a more resilient society. *Environmental Hazards*. doi:10.1080/17 477891.2019.1608148.

Hunt, A., Ferguson, J., Baccini, M. et al. (2017). Climate and weather service provision: Economic appraisal of adaptation to health impacts. *Clim Serv* 7: 78–86. doi:10.1016/J.CLISER.2016.10.004.

Husby, T.G. and Koks, E.E. (2017). Household migration in disaster impact analysis: Incorporating behavioural responses to risk. *Natural Hazards* 87 (1): 287–305. doi:10.1007/s11069-017-2763-0.

Jo, C. (2014). Cost-of-illness studies: Concepts, scopes, and methods. *Clinical and Molecular Hepatology* 20 (4): 327–337. doi:10.3350/cmh.2014.20.4.327.

Jongman, B., Hochrainer-Stigler, S., Feyen, L. et al. (2014). Increasing stress on disaster-risk finance due to large floods. *Nature Climate Change* 4: 264–268. doi:10.1038/nclimate2124.

Joseph, R., Proverbs, D., and Lamond, J. (2015). Assessing the value of intangible benefits of property level flood risk adaptation (PLFRA) measures. *Natural Hazards* 79 (2): 1275–1297.

King, M.A., Dorfman, M.V., Einav, S. et al. (2016). Evacuation of intensive care units during disaster: Learning from the Hurricane Sandy experience. *Disaster Medicine and Public Health Preparedness* 10 (1): 20–27. doi:10.1017/dmp.2015.94.

Kirtman, B., Power, S.B., Adedoyin, J.A. et al. (2013). Near-term Climate Change: Projections and Predictability. Contribution of Working Group I to the Fifth Assessment Report of the Intergovernmental Panel on Climate Change. In: *Climate Change 2013: The Physical Science Basis* (ed. T.F. Stocker, D. Qin, G.-K. Plattner et al.). Cambridge, UK and New York, USA: Cambridge University Press.

Koks, E., Carrera, L., Jonkeren, O. et al. (2016). Regional disaster impact analysis: Comparing input–output and computable general equilibrium models. *Natural Hazards and Earth System Sciences* 16: 1911–1924.

Koks, E. and Thissen, M. (2016). A multiregional impact assessment model for disaster analysis. *Economic Systems Research* 28 (4): 429–449.

Kousky, C. (2014). Informing climate adaptation: A review of the economic costs of natural disasters. *Energy Economics* 46: 576–592. doi:10.1016/J.ENECO.2013.09.029.

Kreibich, H., Bubeck, P., Van Vliet, M. et al. (2015). A review of damage-reducing measures to manage fluvial flood risks in a changing climate. *Mitigation and Adaptation Strategies for Global Change* 20 (6): 967–989.

Kreibich, H., Van Den Bergh, J.C.J.M., Bouwer, L.M. et al. (2014). Costing natural hazards. *Nature Climate Change* 4: 303–306.

Kron, W. (2005). Flood Risk = Hazard • Values • Vulnerability. *Water International* 30 (1): 58–68.

Kunreuther, H., Heal, G., Allen, M. et al. (2013). Risk management and climate change. *Nature Climate Change* 3 (5): 447–450.

Lamond, J., Rose, C., Bhattacharya-Mis, N. et al. (2018). *Evidence Review for Property Flood Resilience Phase 2 Report*. Flood Re and UWE Bristol.

Lamond, J.E. and Joseph, R.D. (2015). Proverbs DG. An exploration of factors affecting the long-term psychological impact and deterioration of mental health in flooded households. *Environmental Research* 140: 325–334. https://doi.org/10.1016/j.envres.2015.04.008.

Larg, A. and Moss, J.R. (2011). Cost-of-Illness studies. *Pharmacoeconomics* 29 (8): 653–671. doi:10.2165/11588380-000000000-00000.

Luechinger, S. and Raschky, P.A. (2009). Valuing flood disasters using the life satisfaction approach. *Journal of Public Economics* 93 (3–4): 620–633.

Mambrey, V., Wermuth, I., and Bose-O'Reilly, S. (2019). Extreme weather events and their impact on the mental health of children and adolescents. *Bundesgesundheitsblatt, Gesundheitsforschung, Gesundheitsschutz* 62 (5): 599–604. doi:10.1007/s00103-019-02937-7.

Mangham, L.J., Hanson, K., and McPake, B. (2008). How to do (or not to do) … Designing a discrete choice experiment for application in a low-income country. *Health Policy and Planning* 24 (2): 151–158. doi:10.1093/heapol/czn047.

Marsh, K., Phillips, C.J., Fordham, R. et al. (2012). Estimating cost-effectiveness in public health: A summary of modelling and valuation methods. *Health Economics Review* 2 (1): 17. doi:10.1186/2191-1991-2-17.

Martinez, G.S., Linares, C., Ayuso, A. et al. (2019). Heat-health action plans in Europe: Challenges ahead and how to tackle them. *Environmental Research* 176: 108548. doi:10.1016/J.ENVRES.2019.108548.

Mas-Colell, A. (1995). *Microeconomic Theory*. New York: Oxford University Press.

Mechler, R. (2016). Reviewing estimates of the economic efficiency of disaster risk management: Opportunities and limitations of using risk-based cost–benefit analysis. *Natural Hazards* 81 (3): 2121–2147.

Meginnis, K., Burton, M., Chan, R. et al. (2018). Strategic bias in discrete choice experiments. *Journal of Environmental Economics and Management*. https://doi.org/10.1016/j.jeem.2018.08.010.

Mulchandani, R., Smith, M., Armstrong, B. et al. (2019). Effect of insurance-related factors on the association between flooding and mental health outcomes. *International Journal of Environmental Research and Public Health* 16 (7): 1174. doi:10.3390/ijerph16071174.

Mulvaney-Day, N.E. (2005). Using willingness to pay to measure family members' preferences in mental health. *Journal of Mental Health Policy and Economics* 8: 71–81.

Munro, A., Kovats, R.S., Rubin, G.J. et al. (2017). Effect of evacuation and displacement on the association between flooding and mental health outcomes: A cross-sectional analysis of UK survey data. *Lancet Planetary Health* 1 (4): e134–e141. doi:10.1016/S2542-5196(17)30047-5.

Neher, C., Duffield, J., Bair, L. et al. (2017). Testing the limits of temporal stability: Willingness to Pay values among Grand Canyon whitewater boaters across decades. *Water Resources Research* 53 (12): 10108–10120. doi:10.1002/2017WR020729.

Odeyemi, I.A. and Nixon, J. (2013). The role and uptake of private health insurance in different health care systems: Are there lessons for developing countries? *ClinicoEconomics and Outcomes Research: CEOR* 5: 109–118. doi:10.2147/CEOR.S40386.

OECD (2018) *Cost-Benefit Analysis and the Environment*, doi:doi:https://doi.org/10.1787/9789264085169-en

Okuyama, Y. and Santos, J.R. (2014). Disaster impact and input–output analysis. *Economic Systems Research* 26 (1): 1–12. doi:10.1080/09535314.2013.871505.

Paranjothy, S., Gallacher, J., and Amlot, R. (2011). Psychosocial impact of the summer 2007 floods in England. *BMC Public Health* 11. doi:10.1186/1471-2458-11-145.

Peters, K. (2018). *Accelerating Sendai Framework Implementation in Asia Disaster Risk Reduction in Contexts of Violence, Conflict and Fragility*. https://cdn.odi.org/media/documents/12284.pdf Accessed 6 September, 2021.

Phalkey, R.K. and Louis, V.R. (2016). Two hot to handle: How do we manage the simultaneous impacts of climate change and natural disasters on human health? *The European Physical Journal. Special Topics* 225: 443–457. https://link.springer.com/article/10.1140/epjst/e2016-60071-y.

Poussin, J.K., Botzen, W.J.W., and Aerts, J.C.J.H. (2015). Effectiveness of flood damage mitigation measures: Empirical evidence from French flood disasters. *Global Environmental Change* 31: 74–84.

Powdthavee, N. and Van Den Bergh, B. (2011). Putting different price tags on the same health conditions: Re-evolving the well-being evaluation approach. *Journal of Health Economics* 30 (5): 1032–1043.

Prettenthaler, F., Kortschak, H.S.S., Mechler, R. et al. (2015). Catastrophe management: Riverine flooding. In: *Economic Evaluation of Climate Change Impacts* (ed. K.W. Steininger, M. Koning, B. Bednar-Friedl, et al.), 349–366. Springer International Publishing.

Rakotonarivo, O.S., Schaafsma, M., and Hockley, N. (2016). A systematic review of the reliability and validity of discrete choice experiments in valuing non-market environmental goods. *Journal of Environmental Management* 183: 98–109. https://doi.org/10.1016/j. jenvman.2016.08.032.

Roschnik, S., Martinez, G., Yglesias-Gonzalez, M. et al. (2017). Transitioning to environmentally sustainable health systems: The example of the NHS in England. *Public Health Panoor* 3 (2): 229–236.

Rosenheim, N., Grabich, S., and Horney, J.A. (2018). Disaster impacts on cost and utilization of Medicare. *BMC Health Services Research* 18 (1): 89. https://doi.org/10.1186/s12913-018-2900-9.

Ryan, M., Mentzakis, E., Jareinpituk, S. et al. (2017). External validity of contingent valuation: Comparing hypothetical and actual payments. *Health Economics* 26 (11): 1467–1473. doi:10.1002/hec.3436.

Saulnier, D.D., Brolin Ribacke, K., and Von Schreeb, J. (2017). No calm after the storm: A systematic review of human health following flood and storm disasters. *Prehospital and Disaster Medicine* 32 (5): 568–579. doi:10.1017/S1049023X17006574.

Schmitt, H.M.L., Graham, M.H., and White, C.L.P. (2016). Economic evaluations of the health impacts of weather-related extreme events: A scoping review. *International Journal of Environmental Research and Public Health* 13.

Sekulova, F. and van den Bergh, J.C.J.M. (2016). Floods and happiness: Empirical evidence from Bulgaria. *Ecological Economics* 126: 51–57. https://doi.org/10.1016/j.ecolecon.2016.02.014

Soeteman, L., van Exel, J., and Bobinac, A. (2017). The impact of the design of payment scales on the willingness to pay for health gains. *Euoupean Journal. of Health Economics* 18: 743–760. https://doi.org/10.1007/s10198-016-0825-y

Surminski, S., Aerts, J.C.J.H., Botzen, W.J.W. et al. (2015). Reflection on the current debate on how to link flood insurance and disaster risk reduction in the European Union. *Natural Hazards* 79 (3): 1451–1479. https://doi.org/10.1007/s11069-015-1823-5

Surminski, S. and Thieken, A.H. (2017). Promoting flood risk reduction: The role of insurance in Germany and England. *Earth's Future* 5: 979–1001. https://doi.org/10.1002/2017EF000587

Tarricone R. (2006). Cost-of-illness analysis. What room in health economics? *Health Policy* 77 (1): 51–63. doi: 10.1016/j.healthpol.2005.07.016. Epub 2005 Sep 1. PMID: 16139925

Thieken, A.H., Müller, M., Kreibich, H. et al. (2005). Flood damage and influencing factors: New insights from the August 2002 flood in Germany. *Water Resources Research* 41 (12): W12430. https://doi.org/101029/2005WR004177

Trautmann, S., Rehm, J., and Wittchen, H.U. (2016). The economic costs of mental disorders: Do our societies react appropriately to the burden of mental disorders?. *EMBO reports* 17 (9): 1245–1249. https://doi.org/10.15252/embr.201642951

UNDRR (2015). Sendai framework for disaster risk reduction 2015–2030. 32.

UNEP (2019). UNEP annual report, UNEP, Geneva. https://www.unep.org/annualreport/2019/index.php

United Nations (1992). United Nations framework convention on climate change. 33.

United Nations (2015). Transforming our world: The 2030 Agenda for sustainable development. 35.

Unterberger, C., Hudson, P., Botzen, W.J.W. et al. (2019). Future public sector flood risk and risk sharing arrangements: An assessment for Austria. *Ecological Economics* 156: 153–163, https://doi.org/10.1016/j.ecolecon.2018.09.019

von Möllendorff, C. and Hirschfeld, J. (2016). Measuring impacts of extreme weather events using the life satisfaction approach. *Ecological Economics* 121. https://doi.org/10.1016/j.ecolecon.2015.11.013

Waite, T.D., Chaintarli, K., Beck, C.R. et al. (2017). The English national cohort study of flooding and health: Cross-sectional analysis of mental health outcomes at year one. *BMC Public Health* 17: 129. https://doi.org/10.1186/s12889-016-4000-2

Watts, N. (2018). The 2018 report of the Lancet Countdown on health and climate change: Shaping the health of nations for centuries to come. *The Lancet* 392 (10163): 2479–2514.

WHO (2003). *Investing in Mental Health*. Geneva, Switzerland: World Health Organization.

WHO (2008). *Heat-Health Action Plans: A Guidance*. (Matthies, F., Bickler, G., Cardeñose A., et al. eds.). Copenhagan, Denmark: World Health Organization Regional Office for Europe.

WHO (2012). *Atlas of Health and Climate*. Geneva, Switzerland: World Health Organization.

WHO (2013). *Floods in the WHO European Region: Health Effects and Their Prevention*. Copenhagan, Denmark: World Health Organization Regional Office for Europe.

WHO (2015a). *Operational Framework for Building Climate Resilient Health Systems*. Geneva, Switzerland: World Health Organization.

WHO (2015b). *Comprehensive Safe Hospital Framework*. Geneva, Switzerland: World Health Organization.

WHO (2015c). *Hospital Safety Index Guide fpr Evaluators*. Geneva, Switzerland: World Health Organization.

WHO (2018). *Factsheet: Climate Change and Health*. Geneva, Switzerland: World Health Organization.

Winsemius, H.C., Aerts, J.C.J.H., van Beek, L.P.H. et al. (2016). Global drivers of future river flood risk. *Nature Climate Change* 6 (4), 381–385. https://doi.org/10.1038/nclimate2893

Zhong, S., Yang, L., Toloo, S. et al. (2018). The long-term physical and psychological health impacts of flooding: A systematic mapping. *Science of The Total Environment* 626. https://doi.org/10.1016/j.scitotenv.2018.01.041.

8

Conclusions and Perspectives

Franziska Matthies-Wiesler[1] and Philippe Quevauviller[2]

[1]*Institute of Epidemiology, Helmholtz Zentrum München, German Research Center for Environmental Health*
[2]*Vrije Universiteit Brussel (VUB)*

8.1 Climate Change Mitigation Vs. Adaptation

The Paris Agreement, adopted in 2015, aims to keep global warming well below 2 °C, or even better, limit it to 1.5 °C (UN 2015a). In order to achieve this goal, CO_2 emissions need to be reduced to reach net zero by 2050. In addition, the agreement aims to increase climate change adaptation of parties. Ambitious global, national and local climate change mitigation measures across sectors are required to reduce CO_2 emissions and limit global warming to the set goal. To reach this goal, Member States have set specific nationally determined contributions (NDCs) (United Nations Climate Change 2019). Climate change mitigation measures potentially carry health co-benefits, not least by reducing the magnitude of climate change and its respective health effects, but also the measures themselves. Examples include a transformation of the food system to a sustainable system that respects the planetary boundaries and of the fossil fuel-based transport system towards alternative modes of transport and active mobility.

Climate change mitigation would also limit or reduce the risks from hydrometeorological extreme events, by reducing the risk of droughts and the risks of flooding, e.g., from sea-level rise related to the melting of glaciers and ice sheets and to expansion of sea water as it warms (Stocker et al. 2013). Chapters 2, 3, 4 and 7 outlined the projected risks of hydrometeorological extreme events in relation to climate change and also the respective health impacts and costs. Climate change mitigation measures with large leaverage include the changes towards sustainability and zero emissions in the energy, the transport, the building and the health sector, as well as changes towards sustainable food system (Springmann et al. 2018) and land use changes (German Advisory Council on Global Change (WBGU) 2021). Globally, the healthcare sector is responsible for 4.6% of the CO_2 emissions, thus the sector represents one of the strong levers in climate change mitigation (Watts et al. 2019).

Even if climate change is being limited through effective mitigation measures, adaptation to the changes already occurring now and in the nearer future will have to be implemented in parallel. For the areas of hydrometeorological extreme events and health, these measures expecially focus on early warning, preparedness and response, as described and discussed in Chapters 4, 5 and 6. Multi-risk governance frameworks related to climate extremes, shifting from single to multi-risk thinking in governmental agencies, represents the key challenge for the future, considering how measures to improve the resilience of the built environment and communities may provide effective solutions to strengthen adaptation measures. This is reflected in climate-related research orientations of the Horizon Europe Framework Programme (European Commission 2021b).

It is important to ensure that adaptation measures do not exacerbate climate change and pose a risk to health. Health and environmental impact assessment need to be part of the development process. Potentially, adapatation measures can be designed to contribute to climate change mitigation and environmental protection, e.g., when reconsidering land use, selecting crops or developing technological improvements for water management (WBGU 2021).

There are limits to adaptation to climate change: weak health systems in countries with low and medium income will only be able to cope with the requirements and the adapatation needs to a certain extent. Also, indiviudals with lower socio-economic status often are more exposed to climate change-related health risks and have less options for and access to adaptation. They bear the higher health risks. In general, stronger adaptation efforts will need to be undertaken as climate change progresses; however, it will become increasingly difficult to absorb the increasing risks to human health and livelihoods, and limits to adaptation will be reached (Figure 8.1).

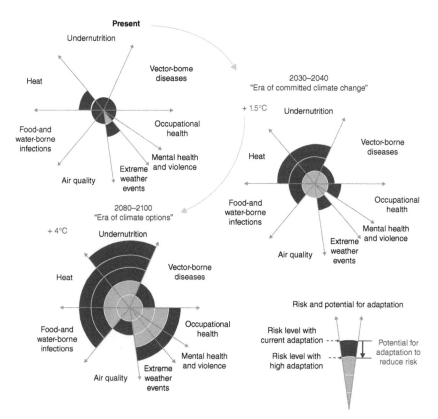

Figure 8.1   Possibilities and limits to climate change adaptation for selected health risks for observed climate change and a 4 degrees-world in 2080–2100 (Smith et al. 2014).

The Sendai Framework for Disaster Risk Reduction aims to reduce disaster risk through prevention and reduction of exposure and increase in preparedness for reponse and recovery and argues for building back better (UNDRR 2015). When looking at the current global health crises caused by the COVID-19 pandemic, building better, greener and fairer in terms of protecting the climate, biodiversity and the planet as a whole would contribute to safeguarding our natural resources (WHO 2020). Looking at the challenging crises of our times (climate change, the environmental, the health and the social crises), reaching the sustainable development goals (SDGs) of the United Nations (UN 2015b), capitalising on their synergies and working towards solving the trade-offs is more urgent than ever (Independent Group of Scientists appointed by the Secretary-General 2019). A Health in all Policies approach and aiming at universal health coverage (UHC) and at leaving no one behind are important health concepts that influence a range of SDGs and their targets, beyond the specific health target (SDG 3). A mind shift towards a more holistic planetary approach is needed to tackle these tasks (Whitmee et al. 2015). Solutions may materialise when identifying and addressing common drivers of environmental changes and ill health.

8.2 Solution-Oriented Research

To address the described climate, environment and health challenges, transdisciplinary and participatory (citizens, authorities, practioners and various stakeholders) research, focusing on the most vulnerable, is needed to develop solutions and to investigate pathways to implementation (e.g., Ebi et al.; HERA Consortium, 2021; Frumkin 2015). In terms of research topics, advances have been reached on early warnings using impact forecasts (e.g., the ANYWHERE project[1]); these need to be expanded to also project health impacts (e.g., for heatwaves), connecting the modelling and the health research communitites. The development of multi-hazard disaster preparedness and response plans is required, together with strategies to ensure their implementation on national, regional and local levels, be it through regulation or incentives. Examples of solution-oriented research developments are taken from the Horizon Europe Work Programme, in particular the Civil Security for Society's Disaster-Resilient Societies area (DRS). With regard to citizens and communities, it is necessary to design preparedness actions enabling an empowerment of citizens (including particularly vulnerable groups), and their communities and NGOs through bottom-up participatory and learning processes. This includes school/university curricula and professional training and trust building amongst local actors, integrating relevant traditional knowledge, incorporating a gender perspective where relevant, best practices, guidelines, and possible changes of regulations, to allow participatory actions. Difficulties in communication to the public in preparedness (and response) phases requires the consideration of legal aspects, along with investigations into innovative practices, forms and tools that will enable the more effective sharing of information, taking into account possible risks of disinformation and fake news (European Commission 2021b). Improved Disaster Risk Management (DRM) and governance is another important feature requiring research support. In this area, a focus is made on Integrated Disaster Risk Reduction for extreme climate events which includes some relevant components for pandemics, in particular in addressing the capacity of communities and governments to manage expected and/or unexpected extreme climate events. The effective governance throughout the entire DRM cycle indeed also covers risks of pandemics (European Commission 2021b).

Health impacts of droughts, desertification and land degradation are more difficult to assess as they are longer onset impacts and partly more complex. Even migration and conflict are potential consequences of climate change-induced hydrometeorological extreme events and the interlinkages require more attention in research (e.g., Ghimire et al. 2015; Abel et al. 2019).

Identifying and addressing common drivers of climate change, environmental degradation and ill health can have knock-on effects and co-benefits across these areas (HERA Consortium, 2021). For the required systemic thinking, scientific, sectoral, social and geographical silos need to be overcome. Integrated approaches are essential to bridge different policy areas including civil protection, environment (including water, forestry, biodiversity/nature), climate change adaptation and mitigation, health and consumer protection, and security.

In the societal resilience domain, new topics include efforts towards a better understanding of citizens' behavioural and psychological reactions in the event of a disaster or crisis situation (European Commission 2021b). To facilitate and accelerate the implementation of co-designed solutions, research into transformational change is needed, such as what are the drivers and what are the barriers of change and how can they be overcome? (HERA Consortium, 2021). How can transformational change be triggered – how much bottom-up and how much top-down initiation is needed to reach tipping points (Otto et al. 2020)? The concept of planetary health, in research and the development of policies, means a change of paradigm with regard to the relationship between human health, economy and nature, as 'human health and human civilisation depend on flourishing natural systems and the wise stewardship of those natural systems' (Whitmee et al. 2015).

8.3   Community-Building

Research project outputs in many sectors, including climate and health sciences, materialise in many different ways, such as relevant scientific findings, the maturation of promising technology areas, the operational validation of innovative concepts or the support to policy implementation. Over the years, a strong security research community, driven forward by highly committed stakeholders, has been consolidated at different levels (international, European and national). In the climate and health areas, policymakers, practitioners, industry, scientists and citizens (civil society, municipalities, etc.) are the pillars on which sound research agendas are built. The dialogue amongst different actors guarantees not only that research addresses real needs, but also that the investment in research will deliver tangible results. One example of a wide community-building aiming to facilitate interactions within the research community and users in various areas of security research is the Community of Users for Safe, Secure and Resilient Societies (CoU) launched by the European Commission in 2014. It includes the Disaster-Resilient Societies (DRS) Thematic Area, covering issues related to HEEs and their impacts on health. This informal platform enabled to gather policy-makers, scientists, practitioners, industry/SMEs and civil society organisations at international and regional level, creating dialogues around research in various thematic areas and building 'bridges' amongst different sectors (areas, disciplines and actors), which was an essential step forward. Dialogues and events had a clear effect on enhancing the participation of practitioners in research projects, in particular by promoting research results that are relevant to them, including the most promising tools that might have the potential to be taken up by them, and ensuring that their expertise is made available to policy-makers. Synergies were also stimulated between research and capacity-building projects. In the light of the positive impacts of this community-building, the initiative has grown up into a more ambitious network of networks in support of Horizon Europe: the so-called Community for European Research and Innovation for Security (CERIS). This forum has a large scope with various thematic areas, with the following objectives:

- Raising awareness on major updates in relevant policy sectors and on results achieved by related research and non-research initiatives, analyse impacts and provide policy recommendations.
- Analysing identified capability needs and gaps in the corresponding thematic areas (within Thematic Working Groups and other networks) and prioritisation of related research orientations based, at least, on criticality and urgency, in order to produce recommendations for a civil security research agenda.
- Identifying solutions available to address the gaps, differentiating state-of-the-art technologies (off-the-shelf and development and integration products) and security research trends. It will also take into account other considerations, such as technological maturity, operational relevance, societal acceptance, cost-effectiveness, etc.
- Translating capability gaps and potential solutions into research needs (including scenarios linking research needs to capabilities and societal appropriation, Technology Readiness Levels, development roadmaps, research action types, perspectives of research uptake, etc.) and get feedback from practitioners about prioritisation of the needs, inputs to research programming and involvement in research activities.
- Identifying funding opportunities and synergies between different funding instruments, and proposing measures to facilitate them.
- Identifying standardisation needs through existing networks/platforms and prioritise them in close consultations with policy-makers and practitioners.
- Integrating the views of citizens so as to promote responsible research and innovation, which respects ethical considerations and civil liberties.

8.4 Strengthening International Partnerships

Our societies nowadays have to deal with complex and transboundary crises within which a more systemic approach with strict interconnection between risk reduction and sustainable development is needed. Partnerships and collaboration are mandatory to jointly address these challenges, in research, policy development and their implementation (UN 2016).

Risk reduction of any kind of disasters is regulated by a number of international, EU and national and local policies and strategies covering various sectors and features, such as awareness raising, prevention, mitigation, preparedness, monitoring and detection, response and recovery. In this respect, the implementation of international policy initiatives (e.g., the Sendai Framework for Disaster Risk Reduction) requires cross-border and cross-sectoral cooperation as well as enhanced collaboration amongst different actors and strengthened knowledge covering the whole disaster management cycle, from prevention and preparedness to response and recovery (and learning).

Understanding and exploiting the existing linkages and synergies amongst policy initiatives such as the Paris Agreement (UN 2015a), the EU strategy on adaptation to climate change (European Commission 2021a), the EU Green Deal (European Commission 2019), the Sendai Framework (UNDRR 2015) and the Sustainable Development Goals (SDGs) (UN 2015b) represents in this sense a global priority for future research and innovation actions in the field of natural hazards and man-made disasters.

For the response side, international cooperation on research and innovation with key partners has the potential to identify common solutions and increase the relevance of outcomes. As such, the International Forum to Advance First Responder Innovation (IFAFRI) and other Expert Networks involved in UN and/or NATO initiatives, for example, have provided overviews of existing gaps and are in the position to engage in cooperation with partners inside and outside the EU, the results of which can provide a valuable source for identifying most urgent needs concerning disaster management (e.g., knowledge, operational, organisational and technological) of relevance to international cooperation, in particular in support of the implementation of international policies such as the Sendai Framework for Disaster Risk Reduction (UNDRR 2015).

Enhanced cooperation and involvement of different sectors and actors are essential, including policy-makers, scientists from various disciplines, industry/Small and Medium Enterprises (SMEs), public administration (both at national and regional/local level), credit/financial institutions, practitioners, Non-Governmental Organisations (NGOs) and Civil-Society Organisations (CSOs), notwithstanding the citizen dimension. Cross-regional, cross-border and cross-sector agreements covering all phases of DRM can improve the knowledge about extreme climate events such as forest fires, droughts, floods, heatwaves, storms and storm surges (European Commission 2021b).

In addition, improving effective prevention, preparedness and response rely upon specific national or local expertise and experience. Like in research, it is important to overcome silos between technical and political authorities at all levels and advocate integration amongst involved actors.

Social inequality between countries and within exacerbate impacts of climate change and environmental degradation. Mitigation and adaptation measures need to focus on the most vulnerable population groups (socio-economic factors) and vulnerable countries. Social inequalities affect most aspects of life, air quality, housing, noise, occupational health, water and sanitation and the impacts of climate change.

In the set of sustainable development goals, strengthening the international and global partnership (SDG 17) is important for sustainable development and for reaching the overall goal as well as individual targets (UN 2015b). While the COVID-19 pandemic has caused a decline in overseas development aid to middle- and low-income countries in 2020 (United Nations and Department of Economic and Social Affairs 2021), the distribution of vaccines and medical appliances has even more stirred the discussion on international responsibility and partnership.

In an analysis of positive and negative interrelations between the SDGs, overall positive effects could be identified for SDG 17 (Table 8.1). Partnerships and cooperation are effective pathways to support sustainable development for all ('leaving no one behind') and correspond to the global responsibility for solutions to address major global crises.

Table 8.1 Positive and negative interrelations between the 17 SDGs (Independent Group of Scientists appointed by the Secretary-General 2019).

Sum of active influence: horizontal; sum of being influenced: vertical. Positive effect = blue; negative effect = orange; interrelations for which no information is available, are represented through empty boxes.

Sustainable development in turn has great potential for transformative change (Independent Group of Scientists appointed by the Secretary-General 2019). However, human capabilities need to be strengthened to build capacity for increased resilience, sustainable development and transformational change (Figure 8.2).

Education, the satisfaction of basic needs (e.g., food, water, shelter and healthcare) and access to information and services and empowerment are crucial to reduce social inequality, to increase resilience and capabilities.

8.5 Perspectives

Enhanced risks related to HEEs and their impacts on health call for enhanced capacities in risk and resilience management and governance, including instruments for better prevention and preparedness, technologies for operational practitioners, and where relevant for citizens, and overall for societal resilience. The increasing severity and frequency of extreme weather events (e.g., floods, heat and cold waves, storms, forest fires) combined with health-related crises such as the

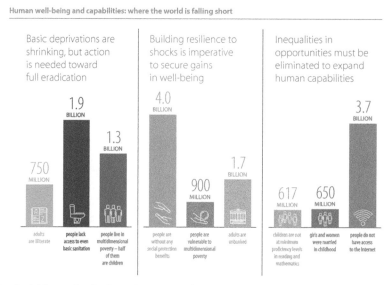

Figure 8.2   Social inequality (Independent Group of Scientists appointed by the Secretary-General 2019).

COVID-19 pandemic, have demonstrated how societies have become more exposed and vulnerable to pandemic risks as existing global inequalities often exacerbate both the exposure and vulnerability of communities, infrastructures, health systems, economies and nature. In fact, individuals most vulnerable to COVID-19 are also most at risk for health impacts from heat – and other extreme events. (Bose-O'Reilly et al. 2021) Climate change and environmental degradation (including biodiversity loss) increase the global risk of pandemics (IPBES, 2020). Mitigating these environmental changes, including climate change, e.g., through directing recovery efforts (such as financial incentives and support) towards sustainability (greener, fairer and healthier), contributes to improving health and to reducing the risk of pandemics in the future (WHO 2020). In addition to strengthening health systems, preparedness and response, a broad range of short-term and longer-term research needs has been identified in this nexus (Destoumieux-Garzon et al. 2021; Barouki et al. 2021).

Amongst other needs, it is essential to address the common roots and causes of environmental and climate change-related risks to health, well-being and livelihoods. New research programmes – but also ways towards implementation – are good perspectives. The European Green Deal gives the opportunity for Europe to lead by example and calls for the need to assume international and global responsibility at the same time (European Commission 2019). For example, a coherent integration between Disaster Risk Reduction, Climate Change Adaptation policies and Sustainable Development Goals as fostered by the Green Deal and its respective strategies[2], for example on zero pollution, climate change, energy, circular economy, biodiversity and food systems and UN major initiatives should result in a comprehensive resilience framework, while improving synergies and coherence amongst the institutions and international agencies involved (European Commission 2021b). It is urgent to foster innovative intersectorial solutions-oriented research that is based on solid evidence, feeding into policy, ensuring action and assessing their effects. For this, the disaster, the health and the environment and climate community as well as decision-makers, stakeholders, actors and citizens need to interact to bring together their research ideas and results and experience from practice.

Notes

1. ANYWHERE (anywhere-h2020.eu).
2. A European Green Deal | European Commission (europa.eu).

References

Abel, G.J., Brottrager, M., Cuaresma, J.C. et al. (2019). Climate, conflict and forced migration. *Global Environmental Change*, 54: 239–49.

Barouki, R., Kogevinas, M., Audouze, K. et al. (2021). The COVID-19 pandemic and global environmental change: Emerging research needs. *Environment International*, 146: 106272.

Bose-O'Reilly, S., et al. (2021). COVID-19 and heat waves: New challenges for healthcare systems *Environmental Research*, 198: 111–153.

Destoumieux-Garzon, D., Matthies-Wiesler, F., Bierne, N. et al. (2021). Getting out of crises: Environmental, social-ecological and evolutionary research needed to avoid future risks of pandemics.

Ebi, K. et al. (2020). Transdisciplinary research priorities for human and planetary health in the context of the 2030 Agenda for Sustainable Development, Int. J. Environ. Res. Public Health, 17, 23, 8890 https://doi.org/10.3390/ijerph17238890

European Commission. The European Green Deal. 24.

European Commission (2021a). Forging a climate-resilient Europe – the new EU Strategy on Adaptation to Climate Change. 23.

European Commission (2021b). Horizon Europe Work Programme 2021–2022. *Civil Security for Society*, 6: 214.

Frumkin, H. (2015). Towards consequential environmental epidemiology. Commentary. *Epidemiology*, 26.

German Advisory Council on Global Change (WBGU) (2021). *Rethinking Land in the Anthropocene: From Separation to Integration.* Berlin, Germany.

Ghimire, R., Ferreira, S., and Dorfman, J.H. (2015). Flood-induced displacement and civil conflict. *World Development*, 66: 614–628.

HERA Consortium (2021). EU Research Agenda for the environment, climate & health 2021-2030 – final draft, 86 pages. Available at: https://heraresearcheu.eu?

Independent Group of Scientists appointed by the Secretary-General (2019). *Global Sustainable Development Report 2019: The Future is Now – Science for Achieving Sustainable Development.* New York.

IPBES (2020). *Workshop Report on Biodiversity and Pandemics of the Intergovernmental Platform on Biodiversity and Ecosystem Services* (ed. P. Daszak, J. Amuasi, C.G. Das Neves, et al.). Bonn, Germany: IPBES Secretariat.

Otto, I., Donges, J.F., Cremades, R. et al. (2020). Social tipping dynamics for stabilizing Earth's climate by 2050. *PNAS*, 117: 2354–65.

Smith, K.R., Woodward, A., Campbell-Lendrum, D. et al. (2014). Human health: Impacts, adaptation, and co-benefits. In *Climate Change 2014: Impacts, Adaptation, and Vulnerability. Part A: Global and Sectoral Aspects. Contribution of Working Group II to the Fifth Assessment Report of the Intergovernmental Panel on Climate Change* (ed. C.B. Field, V.R. Barros, D.J. Dokken, et al.). Geneva, Switzerland.

Springmann, M., Clark, M., Mason-D'Croz, D. et al. (2018). Options for keeping the food system within environmental limits. *Nature*, 562: 519–25.

Stocker, T.F., Qin, D., Plattner, G.-K. et al. (2013). Technical Summary. In *Climate Change 2013: The Physical Science Basis*. (Stocker, T.F., Qin, D., Plattner, G.-K., eds.). Geneva, Switzerland: Contribution of Working Group I to the Fifth Assessment Report of the Intergovernmental Panel on Climate Change. Cambridge, UK and New York: Cambridge University Press.

UNDRR (2015). Sendai framework for disaster risk reduction 2015–2030. 32.

United Nations (2015a). Paris Agreement. 27.

United Nations (2015b). Transforming our world: The 2030 Agenda for Sustainable Development. 35.

United Nations (2016). SDG 17; Partnerships: Why they matter. 2.

United Nations and Department of Economic and Social Affairs (2021). Goal 17: Strengthen the means of implementation and revitalize the global partnership for sustainable development.

United Nations Climate Change (2019). Nationally Determined Contributions (NDCs).

Watts, N., Amann, M., Arnell, N. et al. (2019). *The 2019 report of the Lancet Countdown on health and climate change: Ensuring that the health of a child born today is not defined by a changing climate*. London.

Whitmee, S., Haines, A., Beyrer, C. et al. (2015). Safeguarding human health in the Anthropocene epoch: Report of The Rockefeller Foundation–Lancet Commission on planetary health. *Lancet Commission*, 386: 1973–2028.

WHO (2020). WHO manifesto for a healthy recovery from COVID-19: Prescriptions and Actionables for a Healthy and Green Recovery. *World Health Organization*, 1–5. Available at: who-manifesto-for-a-healthy-and-green-post-covid-recovery_4d85f26a-73db-46b7-a2a5-9854ca6faa64.pdf

Index

Hydrometeorological Extreme Events and Public Health, First Edition. Edited by Franziska Matthies-Wiesler and Philippe Quevauviller.
© 2022 John Wiley & Sons Ltd. Published 2022 by John Wiley & Sons Ltd.